PEARSON EDEXCEL GCSE (9–1)
Combined Science

Ali Mclachlan, Peter Robinson,
Helen Tuckey

Pearson

Published by Pearson Education Limited, 80 Strand, London, WC2R ORL.

www.pearsonschoolsandfecolleges.co.uk

Copies of official specifications for all Pearson qualifications may be found on the website: qualifications.pearson.com

Typeset and illustrated by QBS Learning
Produced by QBS Learning

The rights of Ali Mclachlan, Peter Robinson and Helen Tuckey to be identified as authors of this work have been asserted by them in accordance with the Copyright, Designs and Patents Act 1988.

First published 2018

21 20 19 18
10 9 8 7 6 5 4 3 2 1

British Library Cataloguing in Publication Data
A catalogue record for this book is available from the British Library

ISBN 978 1 292 24531 7

Printed in Slovakia by Neografia

Acknowledgements
The authors and publisher would like to thank the following individuals and organisations for their kind permission to reproduce copyright material.

Photographs
(Key: b-bottom; c-centre; l-left; r-right; t-top)
Alamy Stock Photo: Cavallini James/BSIP SA 46; **Shutterstock:** D. Kucharski K. Kucharska 41, 47
All other images © Pearson Education

Note from the publisher

1. While the publishers have made every attempt to ensure that advice on the qualifications and its assessment is accurate, the official specification and associated guidance materials are the only authoritative source of information and should always be referred to for definitive guidance. Pearson examiners have not contributed to any sections in this resource relevant to examination papers for which they have responsibility.
2. Pearson has robust editorial processes, including answer and fact checks, to ensure the accuracy of the content in this publication, and every effort is made to ensure this publication is free of errors. We are, however, only human, and occasionally errors do occur. Pearson is not liable for any misunderstandings that arise as a result of errors in this publication, but it is our priority to ensure that the content is accurate. If you spot an error, please do contact us at resourcescorrections@pearson.com so we can make sure it is corrected.

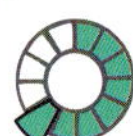

This workbook has been developed using the Pearson Progression Map and Scale for Science.

To find out more about the Progression Scale for Science and to see how it relates to indicative GCSE (9–1) grades go to www.pearsonschools.co.uk/ProgressionServices

Helping you to formulate grade predictions, apply interventions and track progress.

Any reference to indicative grades in the Pearson Target Workbooks and Pearson Progression Services is not to be used as an accurate indicator of how a student will be awarded a grade for their GCSE exams.

You have told us that mapping the Steps from the Pearson Progression Maps to indicative grades will make it simpler for you to accumulate the evidence to formulate your own grade predictions, apply any interventions and track student progress. We're really excited about this work and its potential for helping teachers and students. It is, however, important to understand that this mapping is for guidance only to support teachers' own predictions of progress and is not an accurate predictor of grades.

Our Pearson Progression Scale is criterion referenced. If a student can perform a task or demonstrate a skill, we say they are working at a certain Step according to the criteria. Teachers can mark assessments and issue results with reference to these criteria which do not depend on the wider cohort in any given year. For GCSE exams however, all Awarding Organisations set the grade boundaries with reference to the strength of the cohort in any given year. For more information about how this works please visit: https://qualifications.pearson.com/en/support/support-topics/results-certification/understanding-marks-and-grades.html/Teacher

Contents

Biology

Unit 1 Transport 1
Get started 2
1 How do I explain the factors affecting diffusion rate? 3
2 How do I explain osmosis? 4
3 How do I explain the importance of active transport? 5
Sample response 6
Your turn! 7
Need more practice? 8

Unit 2 Plant biology 9
Get started 10
1 How do I explain how leaf structures are adapted to their functions? 11
2 How do I explain how water is transported in plants? 12
3 How do I explain how limiting factors interact? 13
Sample response 14
Your turn! 15
Need more practice? 16

Unit 3 Genetics 17
Get started 18
1 How do I use genetic terms correctly? 19
2 How do I draw Punnett squares? 20
3 How do I work out probabilities in genetics? 21
Sample response 22
Your turn! 23
Need more practice? 24

Unit 4 Hormones 25
Get started 26
1 How do I explain how hormones control the menstrual cycle? 27
2 How do I explain negative feedback? 28
3 How do I explain the use of hormones in IVF? 29
Sample response 30
Your turn! 31
Need more practice? 32

Unit 5 Genetic modification 33
Get started 34
1 How do I explain the impact of selective breeding? 35
2 How do I describe the stages of genetic engineering? 36
3 How do I explain the benefits and risks of selective breeding and genetic engineering? 37
Sample response 38
Your turn! 39
Need more practice? 40

Unit 6 Calculations in Biology 41
Get started 42
1 How do I do calculations using numbers in standard form? 43
2 How do I convert between units? 44
3 How do I calculate the actual size of very small objects? 45
Sample response 46
Your turn! 47
Need more practice? 48

Unit 7 Answering extended response questions 49
Get started 50
1 How do I know what the question is asking me to do? 51
2 How do I plan my answer? 52
3 How do I choose the right detail to answer the question concisely? 53
Sample response 54
Your turn! 55
Need more practice? 56

Chemistry

Unit 1 Moles 57
Get started 58
1 How do I describe what a mole is? 59
2 How do I calculate how many moles, or particles, there are in a substance? 60
3 How do I calculate empirical formulae? 61
Sample response 62
Your turn! 63
Need more practice? 64

Unit 2 Chemistry calculations 65
Get started 66
1 How do I set out calculations in a logical step-by-step way? 67
2 How do I give answers to an appropriate number of significant figures? 68
3 How do I calculate the mass of a reactant or product? 69
Sample response 70
Your turn! 71
Need more practice? 72

Unit 3 Chemical equations 73
Get started 74
1 How do I balance chemical equations? 75
2 How do I balance an equation given the masses of reactants and products? 76
3 How do I write ionic equations given information about a reaction? 77
Sample response 78
Your turn! 79
Need more practice? 80

Unit 4 Dynamic equilibrium 81
Get started 82
1 How do I describe what dynamic equilibrium means? 83

2 How do I predict changes in equilibrium position caused by temperature changes? 84
3 How do I predict changes in equilibrium position caused by concentration or pressure changes? 85
Sample response 86
Your turn! 87
Need more practice? 88

Unit 5 Energy changes in reactions 89
Get started 90
1 How do I interpret reaction profile diagrams? 91
2 How do I draw reaction profile diagrams? 92
3 How do I calculate energy changes using bond energies? 93
Sample response 94
Your turn! 95
Need more practice? 96

Unit 6 Answering questions about practicals 97
Get started 98
1 How do I describe a suitable experiment? 99
2 How do I change an investigation? 100
3 How do I improve the validity of results? 101
Sample response 102
Your turn! 103
Need more practice? 104

Unit 7 Answering extended response questions 105
Get started 106
1 How do I know what the question is asking me to do? 107
2 How do I plan my answer? 108
3 How do I choose the right detail to answer the question concisely? 109
Sample response 110
Your turn! 111
Need more practice? 112

Physics

Unit 1 Analysing energy transfers 113
Get started 114
1 How do I calculate how much energy is stored in different stores? 115
2 How can I use efficiency to analyse an energy transfer? 116
3 How can I use energy calculations to model events and make predictions? 117
Sample response 118
Your turn! 119
Need more practice? 120

Unit 2 Newton's laws, forces and momentum 121
Get started 122
1 How do I explain acceleration using Newton's laws? 123
2 How do I find the size of the forces causing objects to accelerate? 124
3 How can I describe the momentum of objects? 125
Sample response 126
Your turn! 127
Need more practice? 128

Unit 3 Radioactive decay 129
Get started 130
1 How do I describe the changes that happen during nuclear decay? 131
2 How do I write decay equations? 132
3 How can I use a half-life graph to analyse radioactive decay? 133
Sample response 134
Your turn! 135
Need more practice? 136

Unit 4 Electromagnetism 137
Get started 138
1 How do I relate electricity to magnetism? 139
2 What causes a force to act on a current-carrying conductor? 140
3 How do I calculate the force on a wire? 141
Sample response 142
Your turn! 143
Need more practice? 144

Unit 5 Dealing with equations, calculations and SI units 145
Get started 146
1 How do I choose and use the correct equation? 147
2 How do I know which units to use for quantities? 148
3 How do I set out my calculations to gain full marks? 149
Sample response 150
Your turn! 151
Need more practice? 152

Unit 6 Wave reflection, refraction and absorption 153
Get started 154
1 How do I explain what waves do at boundaries? 155
2 How do I apply ideas about absorption and emission of radiation? 156
3 How do I draw diagrams to explain refraction? 157
Sample response 158
Your turn! 159
Need more practice? 160

Unit 7 Answering extended response questions 161
Get started 162
1 How do I know what the question is asking me to do? 163
2 How do I organise my answer? 164
3 How do I choose the right detail to answer the question concisely? 165
Sample response 166
Your turn! 167
Need more practice? 168

Answers 169

Get started **AO2, AO3**

1 Transport

This unit will help you explain in more detail how diffusion, osmosis and active transport occur. This unit will help you understand the factors that can affect the rate of diffusion.

In the exam, you will be asked to answer questions such as the one below.

Exam-style question

1 A student sets up an experiment to investigate osmosis. He uses the following method:

- make two tubes of Visking tubing (a partially permeable membrane)
- put 25 cm^3 of 20% sucrose solution into tube A
- put 25 cm^3 of pure (distilled) water into tube B
- put both tubes in 5% sucrose solution and leave for 1 hour.

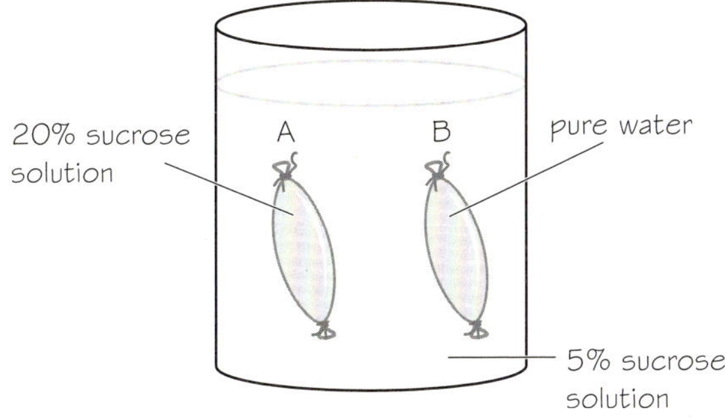
20% sucrose solution
A
B
pure water
5% sucrose solution

After 1 hour, tube B looks smaller than before. Predict the appearance of tube A. Explain your answer. **(4 marks)**

....

You will already have done some work on this topic. Before starting the **skills boosts**, rate your confidence in diffusion, osmosis and active transport. Colour in the bars.

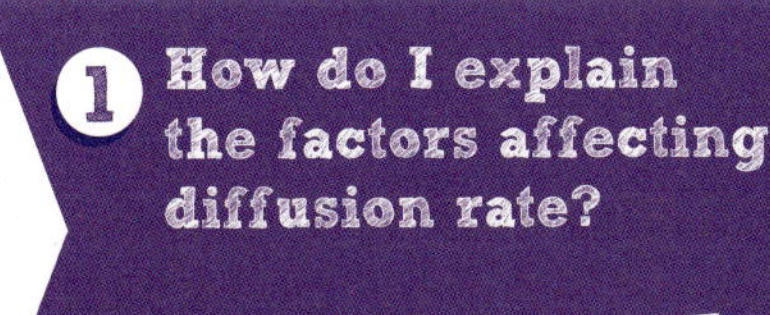
1 How do I explain the factors affecting diffusion rate?

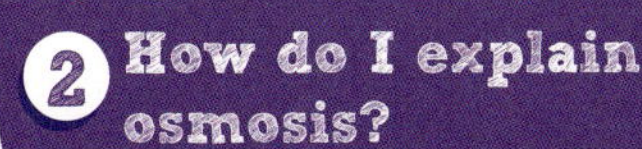
2 How do I explain osmosis?

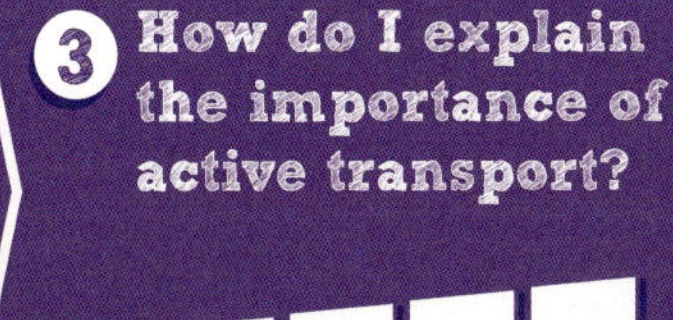
3 How do I explain the importance of active transport?

Biology

Get started

Diffusion is the net movement of substances from an area of high concentration to an area of low concentration. It is a **passive process** (it requires no energy).

Net movement describes the overall amount of movement. If equal amounts of molecules move in opposite directions, there is no overall or net movement.

1. Alveoli in the lungs are where gas exchange occurs in mammals. This is a form of diffusion.

   a. Draw lines to link the gases to the places where they have a higher concentration.

carbon dioxide	higher concentration in alveolus than in capillary
oxygen	higher concentration in capillary than in alveolus

   b. Draw an arrow in blue on the diagram to show the direction in which carbon dioxide molecules will move.

   c. Draw an arrow in red on the diagram to show the direction in which oxygen molecules will move.

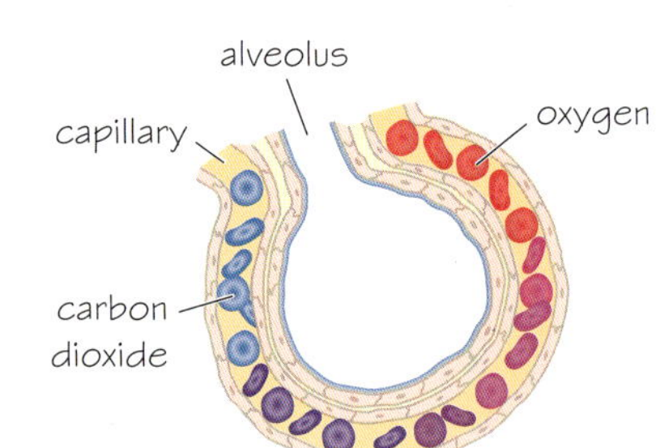

Osmosis is a form of diffusion concerned with the movement of water molecules.

2. Write a definition of osmosis using all the words in the word bank. You can use the words more than once.

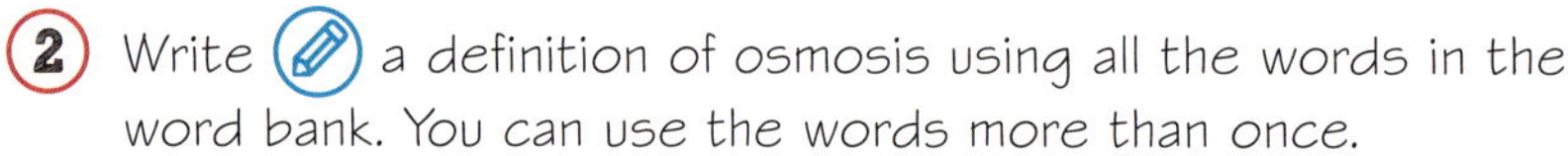

lower	movement	less	region	semi-permeable	concentration
molecules	higher	osmosis	solute	water	membrane	concentrated

....

....

....

3. Diagram 1 shows a model of root hair cells in the roots of plants taking up water by osmosis.

   a. Draw an arrow on diagram 1 to show the direction in which the water molecules will move.

Water molecules move from low to high solute concentration.

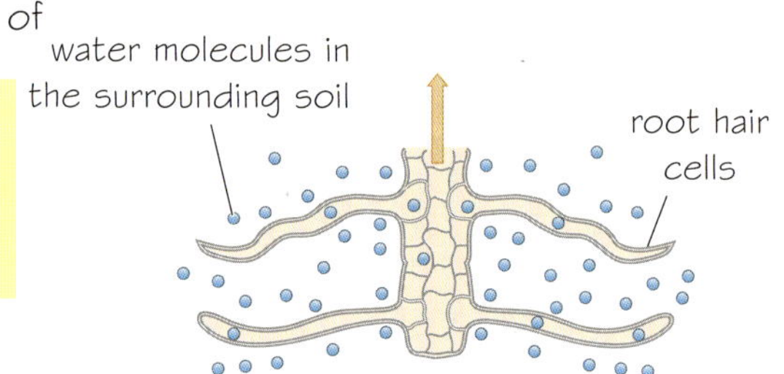

Diagram 1

Active transport is different from diffusion, because it requires **energy**, uses **transport proteins** in the cell membrane and goes **against** a concentration gradient (from low to high).

Root hair cells absorb mineral ions by active transport, as shown in diagram 2.

   b. Draw an arrow on diagram 2 to show the direction in which mineral ions will move.

   c. Circle the correct words in the sentences below to explain how root hair cells absorb mineral ions.

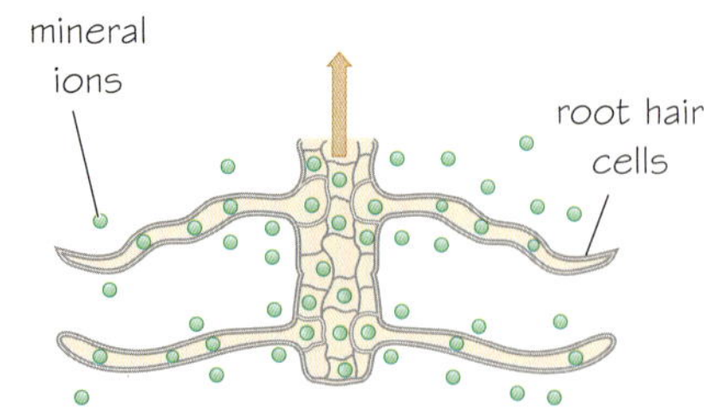

Diagram 2

Mineral ions are at a **higher / lower** concentration in the soil than inside the plant. Root hair cells use **water / energy** to move mineral ions **against / towards** the concentration gradient.

How do I explain the factors affecting diffusion rate?

To help explain how substances are transported into and out of cells by diffusion, it is useful to understand some of the factors that can affect the rate of diffusion.

Rate of diffusion is a measure of the number of randomly moving molecules passing through a surface area per unit time.

Surface area : volume ratio The higher the SA : V ratio, the greater the rate of diffusion.

1 The diagram shows three beakers containing the same volume and concentration of dilute hydrochloric acid. The pink agar blocks contain a mixture of pH indicator and sodium hydroxide. As the acid diffuses into an agar block, it neutralises the sodium hydroxide and turns the block colourless.

A

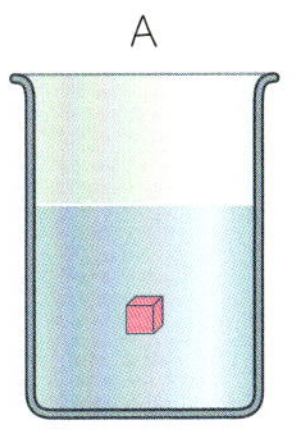

1 cm × 1 cm × 1 cm block of agar

B

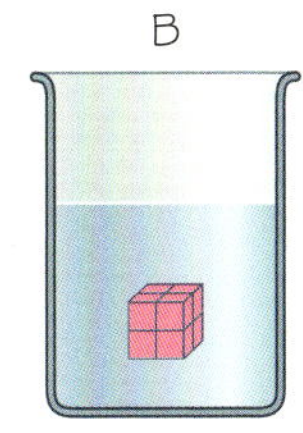

2 cm × 2 cm × 2 cm block of agar

C

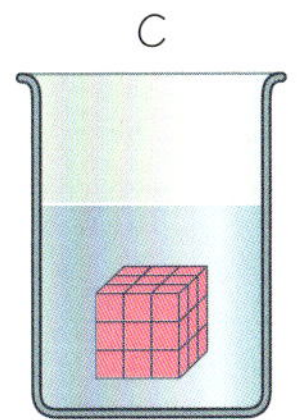

3 cm × 3 cm × 3 cm block of agar

The higher the SA : V ratio, the greater the exposure of the block to the acid.

24 : 8 simplifies to 3 : 1.

6 : 1 is a higher SA : V ratio than 3 : 1.

Complete the table below, using the information to decide the order in which the blocks become colourless.

Block	Surface area (cm^2)	Volume (cm^3)	SA : V	Order of colour change
A	6 (= 1 cm × 1 cm × 6 sides)	1 (1 cm × 1 cm × 1 cm)	6 : 1	
B				

Concentration gradient The greater the difference between the concentrations of substances inside and outside the cell, the faster the rate of diffusion.

2 The diagram shows three identical 2 cm × 2 cm × 2 cm agar blocks. The blocks are placed in the same volume of hydrochloric acid at **different** concentrations.

A

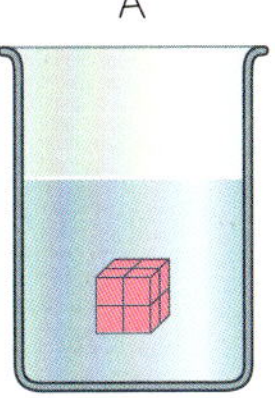

2 cm × 2 cm × 2 cm block of agar

B

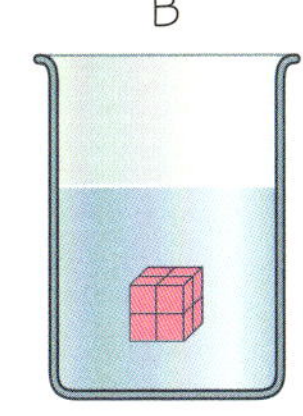

2 cm × 2 cm × 2 cm block of agar

C

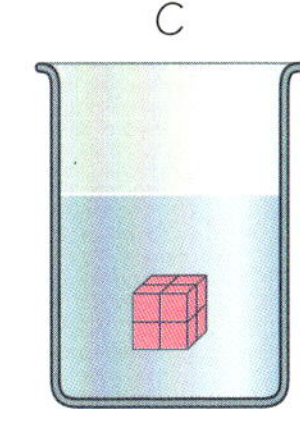

2 cm × 2 cm × 2 cm block of agar

a Complete the table to calculate the rate of diffusion in each beaker.

Beaker	Hydrochloric acid concentration (g/dm^3)	Diffusion path (distance to centre of block)(cm)	Time taken (min)	Rate of diffusion (cm/min)
A	1	1	2	
B	2	1	1	
C	4	1	0.5	

b Complete the following sentences.

The agar block in beaker .... turned colourless first. This is because, compared with other beakers, the .... gradient was .... and the rate of diffusion was the ....

2 How do I explain osmosis?

Explaining osmosis requires an understanding of solute concentration. A solute is a substance dissolved in a solvent such as water. Salt and sugar are common solutes. If a solution has a low solute concentration, then it has a high water concentration. If a cell contains a lower solute concentration than the solution outside, water molecules will move **out of** the cell into the solution, and the cell will shrink.

1. Circle the best words from the words in bold to complete this sentence.

   If a cell contains a much lower solute concentration than in the solution outside, water molecules will move **into** / **out of** the cell, and the cell will eventually **shrink** / **burst**.

2. The diagram shows red blood cells in salt solutions at three different concentrations. Different changes have happened to each red blood cell as a result.

   Indicate whether the solute concentration in each beaker is equal to, higher than or lower than the solute concentration of the cell cytoplasm. Draw lines to match each solution to the correct box.

A

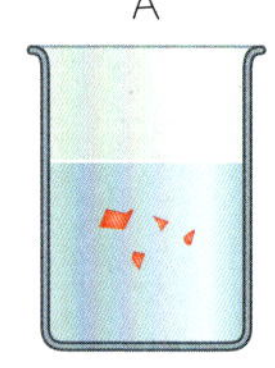

Red blood cell breaks up.

B

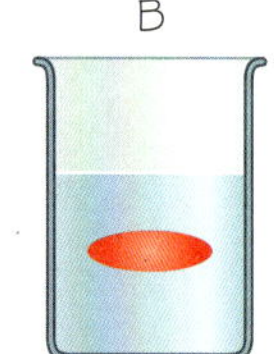

Red blood cell remains the same size.

C

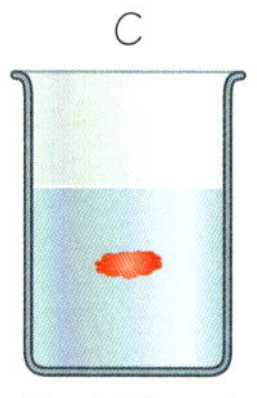

Red blood cell shrivels.

solution in beaker A	equal to
solution in beaker B	higher than
solution in beaker C	lower than

3. Strawberry cells contain water with a low concentration of fruit sugars dissolved in solution. The diagram shows some strawberries with sugar sprinkled on them. After 3 hours, a syrup has formed around the strawberries.

   Explain, in terms of osmosis, why this syrup has formed.

strawberries sprinkled with sugar

strawberries in sugar syrup

....

....

....

What effect does the sprinkled sugar have on the solute concentration outside the strawberry cells?

Remember Water molecules move from a low solute concentration to a higher solute concentration.

4. Three cells of different solute concentrations are placed in a 25 per cent sugar solution, as shown in the diagram.

   Draw arrows from each cell to show the **net** direction of water molecule movement between cells and into or out of the solution, based on the solute concentrations shown.

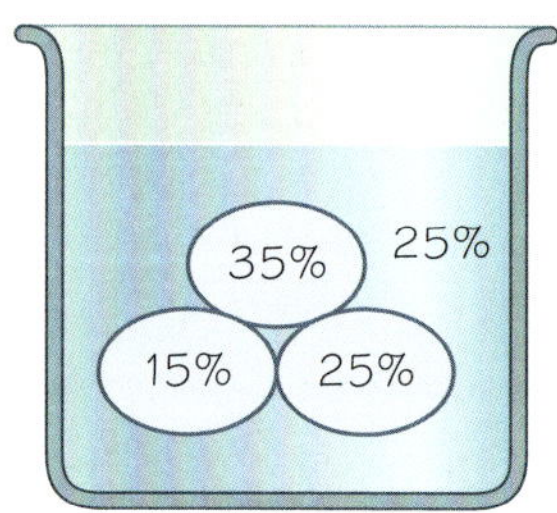

Remember Higher solute concentration = lower water molecule concentration.

How do I explain the importance of active transport?

Unlike diffusion and osmosis, **active transport** requires **energy** to move molecules **against** the concentration gradient through special **transport proteins** in cell membranes.

1 The diagram shows a root hair cell and mineral ions. Active transport is important to plants in order to get low levels of minerals from the surrounding soil into their root system.

Draw lines from the labels to the correct letters (A–E) on the diagram to show how active transport works.

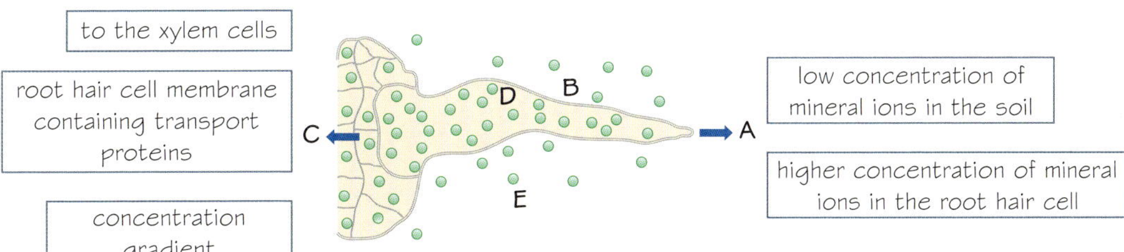

2 In our bodies, active transport plays an important role in digestion. The diagram shows a model of our digestive system. Sodium ions are shown in yellow.

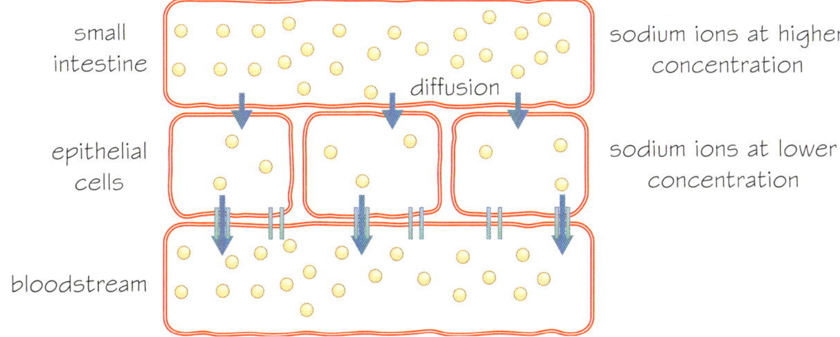

a Circle the correct words in bold to complete this student's explanation of how sodium ions move from the small intestine into the blood.

> Sodium ions are **less / more** concentrated in the small intestine than in the epithelial cells. They move into the epithelial cells by diffusion. Sodium ions are at a **higher / lower** concentration in the epithelial cells than in the bloodstream, and need to move **against / down** the concentration gradient by **osmosis / active transport**. This requires **transport proteins / cellulose** in the cell membrane and **carbon dioxide / energy** from respiration.

b On the diagram below, draw lines to join the labels on the right to the letters A, B and C.

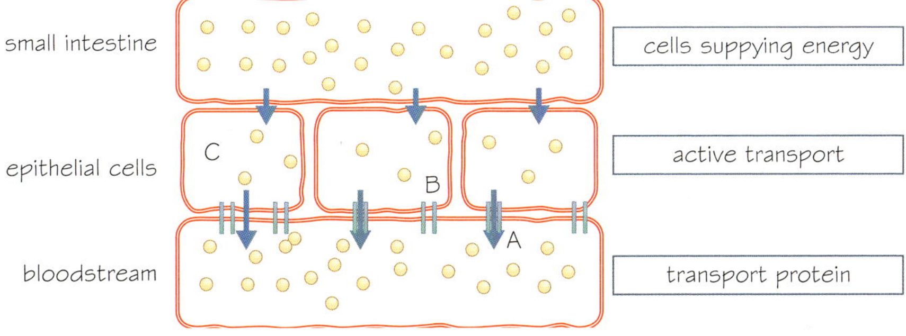

Biology

Sample response

Your understanding of diffusion, osmosis and active transport will often be tested in the context of a core practical. Read this question carefully, use your knowledge and consider your response.

Look at this exam-style question and student response.

Exam-style question

1 A student was investigating osmosis in beetroot cubes. He used the following method:

- Cut beetroot into equal-sized cubes and record the mass of each cube.
- Place each cube into a different concentration of salt solution, then remove the cubes after 30 minutes.
- Pat the beetroot cubes dry with tissue paper, and record the final mass of each cube.

The student's results are shown on the graph below.

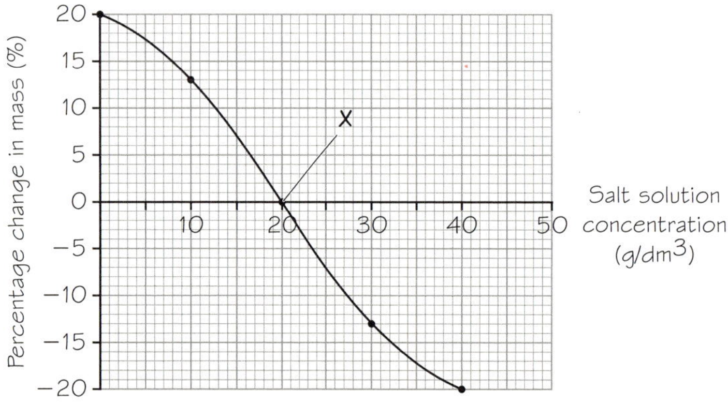

(a) Give a reason why the outside of the beetroot cube must be dried. **(1 mark)**

Water that is still on the cube may add to the mass, so by drying it the cube may have an accurate mass.

(b) Explain the conclusion that can be made about point X on the graph. **(2 marks)**

At a concentration of 20 g/dm^3, there is no change in the mass of the beetroot.

(c) Give one way the student could obtain more data to increase the accuracy of point X. **(1 mark)**

Include a larger range of salt solutions.

1. Justify why the student achieved the mark for **(a)**.

2. The student achieved only 1 mark for **(b)**. How could he have achieved the other mark?

3. The student did not achieve the mark for **(c)**. What response would have achieved the mark? 

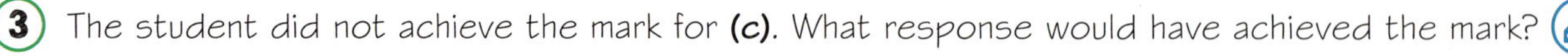

Your turn!

It is now time to use what you have learned to answer this exam-style question.

Remember to read the question thoroughly, looking for information that may help you.

Make good use of your knowledge from other areas of biology.

Exam-style question

1 A student sets up an experiment to investigate osmosis. He uses the following method:

- make two tubes of Visking tubing (a partially permeable membrane)
- put 25 cm^3 of 20% sucrose solution into tube A
- put 25 cm^3 of pure (distilled) water into tube B
- put both tubes in 5% sucrose solution and leave for 1 hour.

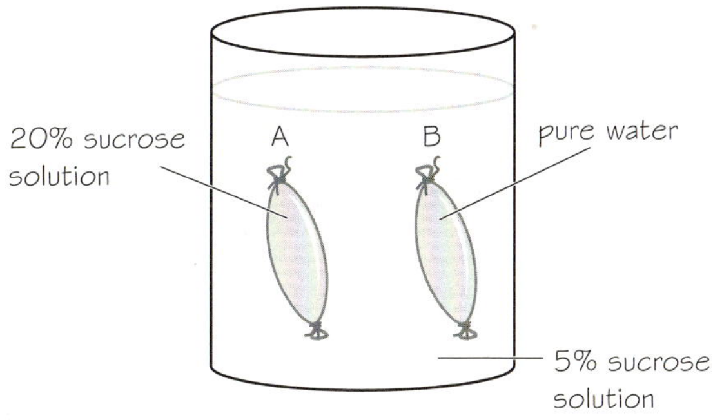

After 1 hour, tube B looks smaller than before. Predict the appearance of tube A. Explain your answer. **(4 marks)**

....

....

....

....

- 'Predict' means say what you think will happen based on what you know.
- There are 4 marks available, so try to identify four different points as answers.
- Which process is taking place: diffusion, osmosis or active transport?
- Will tube A have increased in size, decreased in size, or stayed the same size? Is there a more accurate way of describing this?
- Has the question given you any clues about how the sucrose and water molecules are kept separate?
- Think about where the high and low solute concentrations are and how this will influence the movement of substances.

Need more practice?

Exam questions may ask about different parts of a topic, or parts of more than one topic. Questions about transport of substances could occur as:

- questions about that topic only
- part of a question on how the structure of a tissue or organ is related to its function
- part of a question about an experiment or investigation.

Have a go at these exam-style questions.

Exam-style questions

1 Describe how gases are exchanged through the skin of an earthworm. **(2 marks)**

2 A red blood cell is placed into pure water. The cell bursts. Explain why this happens. **(3 marks)**

3 Explain how mineral ions are absorbed by root hair cells in plants. **(3 marks)**

Boost your grade

To improve your grade, practise answering questions on the osmosis investigation using different examples. You will use potatoes as the example in the lab. However, exam questions may test your ability to apply your knowledge to another example and explain the results of that experiment.

How confident do you feel about each of these **skills**? Colour in the bars.

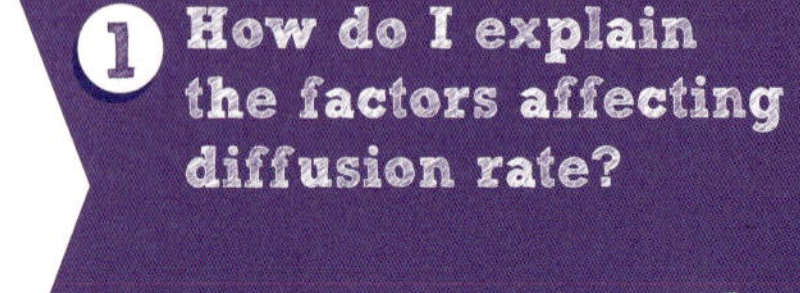

1 How do I explain the factors affecting diffusion rate?

2 How do I explain osmosis?

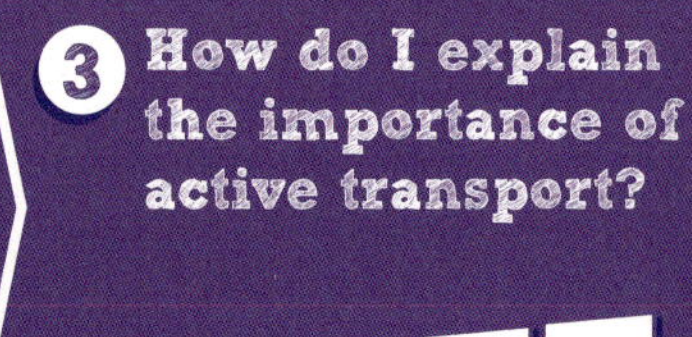

3 How do I explain the importance of active transport?

2 Plant biology

This unit will help you to understand how to explain plant structure and function, and how substances are moved around a plant. It will also help you to understand how factors that limit photosynthesis interact with each other.

In the exam, you will be asked to answer questions such as the one below.

Exam-style question

1 The lower epidermis of a leaf from a bay tree contains many guard cells.

Explain how guard cells help the bay tree survive. **(2 marks)**

....

Figure 1 shows how carbon dioxide concentration affects the rate of photosynthesis in a bay tree.

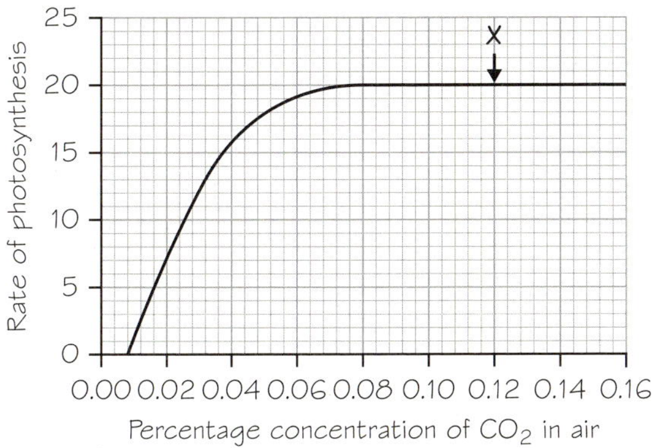

Figure 1

(a) What is the maximum rate of photosynthesis of the bay tree shown by the graph? **(1 mark)**

....

(b) Suggest **one** factor that could be limiting the rate of photosynthesis at point X. **(1 mark)**

....

You will already have done some work on plant biology. Before starting the **skills boosts**, rate your confidence for each skill. Colour in the bars.

1 **How do I explain how leaf structures are adapted to their functions?**

2 **How do I explain how water is transported in plants?**

3 **How do I explain how limiting factors interact?**

Several different types of tissue are found inside a leaf. Each type of tissue is adapted to its function.

1 Using the diagram below, complete the table to show the main types of tissue in the leaf and their functions.

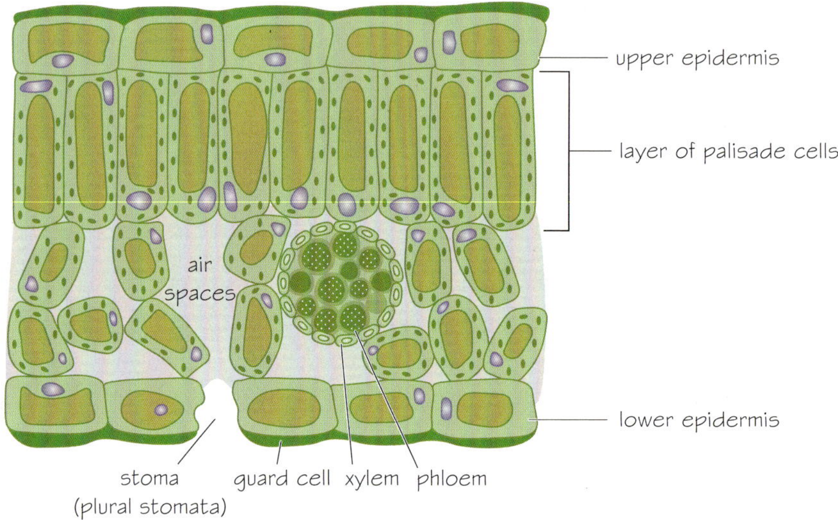

Name of tissue	Description	Function
upper and lower epidermis	covers the outer surface of the leaf contains guard cells that form stomata	protection stomata control gas exchange and water loss
layer of palisade cells	box-shaped cells inside the leaf that contain lots of chloroplasts	
air spaces		provides a large surface area for gas exchange
	dead, hollow tubes strengthened by lignin	transport water and mineral ions from roots to leaves
phloem	tubes made from long cells with pores on the end walls of each cell	

2 Complete the sentences by circling the correct words in **bold** in these sentences.

Water enters the plant through the **root hair cells / guard cells** and is transported by the stem to the leaves in the **phloem / xylem**. Carbon dioxide enters the leaves through the **palisade / epidermis** cells. Dissolved sugar is transported from the leaves around the plant through the **phloem / stomata**.

The rate of photosynthesis is affected by light intensity, carbon dioxide concentration and temperature. If one of these factors becomes too low, it becomes a **limiting factor**, which stops photosynthesis happening any faster.

3 Name **one** factor that is likely to limit the rate of photosynthesis at the top of a mountain.

....

....

To remember the factors that can limit the rate of photosynthesis remember this mnemonic: Leprechauns Cook Toast (Light, Carbon dioxide, Temperature).

 How do I explain how leaf structures are adapted to their functions?

To explain the adaptations of a leaf tissue, you need to link each physical feature to its function. You do this by saying how each feature helps the tissue do its job.

1 Circle the correct words in bold to complete the sentences to describe the physical features of plant epidermal tissue.

> The epidermis covers the surface of the leaf. It is **thick / thin** and **transparent / opaque**. It contains stomata, which **open / close** to allow gas exchange in the **light / dark** for photosynthesis.

Remember A material that is transparent allows light through. A material that is opaque does not allow light through.

2 A student explains how the tissue containing the air spaces is adapted to its function.

> *The tissue containing the air spaces has many irregularly shaped cells with gaps between each one. This provides a large surface area to allow gases to diffuse in and out of its cells easily.*

Think about how the physical features help the tissue carry out its function. Use linking words such as '**to allow**' or '**so that**' or '**because**' to link the physical features to the explanation.

a Highlight the words that give the physical description.

b Circle the words that link the physical description to the explanation.

c Underline the words that explain how the physical feature helps the tissue to carry out its function.

3 Draw lines to link each physical feature of the layer of palisade cells to its correct linking word and explanation of the adaptation.
Not all of the physical features, linking words or explanations need to be used.

A 'palisade' is the name given to a series of wooden stakes arranged in a tightly packed row.

Physical feature of the palisade layer	Linking word	Explanation of the adaptation
The cells contain many chloroplasts	so	they can efficiently exchange gases.
They have a box-like shape	because	the cells can be tightly packed together to fit in as many as possible.
	to	absorb light for photosynthesis.

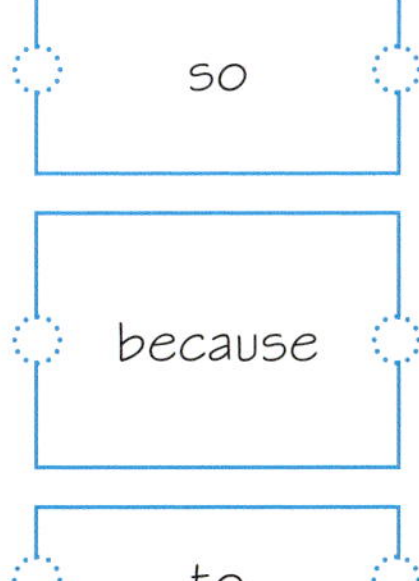

Biology

How do I explain how water is transported in plants?

The movement of water from the roots to the leaves is called **transpiration**. Transpiration explains how water moves up the plant **against gravity** without the use of a pump.

The best way to understand transpiration is to break it down into steps.

1 Complete the flow diagram by numbering the statements in the correct order to show how water is lost from a leaf.

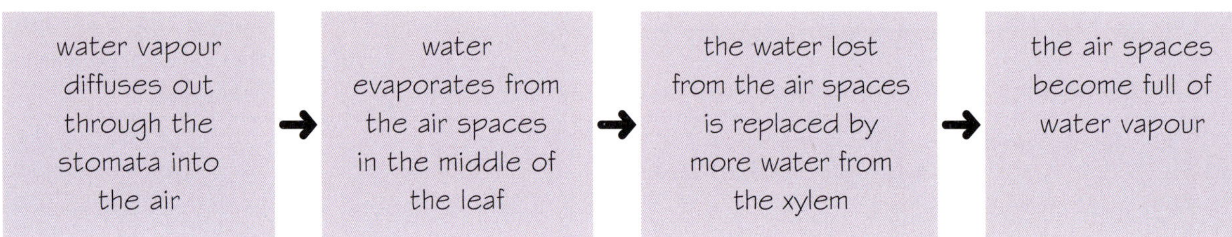

water vapour diffuses out through the stomata into the air
water evaporates from the air spaces in the middle of the leaf
the water lost from the air spaces is replaced by more water from the xylem
the air spaces become full of water vapour

....

2 Describe what happens to water in the root hair cells as water moves up the xylem tissues.

Remember When you are describing the movement of substances, use words that describe their method of transport. Water moves into the root hair cells by **osmosis**.

....

....

Remember HALT: **H**umidity, **A**ir movement, **L**ight and **T**emperature.

Different environmental factors affect transpiration. These include humidity, air movement, light intensity and temperature.

If a factor increases the rate of evaporation and diffusion, it will increase the rate of transpiration.

3 Complete the table.

Change in factor	Effect on transpiration rate	Explanation
humidity increased		The greater the humidity of the air, the more water vapour it contains, so evaporation and diffusion of water from the leaf is slower.
air movement increased		As the rate of air movement increases, water vapour is moved away from the stomata faster. This increases the concentration gradient of water from inside the leaf to outside the leaf, so water ....
light intensity increased		Light causes stomata to open more, allowing ....
temperature increased		The higher the temperature, the more energy water particles have and the faster they move, so water ....

4 At night, water moves out of the guard cells by osmosis. Explain how this can help prevent the plant from wilting. Include the words in the box in your answer.

water	vapour
escape	wilt

When water moves out of the guard cells they become flaccid, causing the stoma ....

....

3 How do I explain how limiting factors interact?

For graphs showing how rates of photosynthesis are affected by a given factor, you need to be able to:

- explain the shape of the graphs for different factors
- understand graphs showing the effect of more than one factor
- identify limiting factors.

The straight line in the first part of the graph shows that the rate of photosynthesis increases in relation to the light intensity. It is a linear relationship.

When the line levels off it is showing that the rate of photosynthesis has stopped increasing.

1 The graph below shows how light intensity affects the rate of photosynthesis. Add the correct labels to the graph.

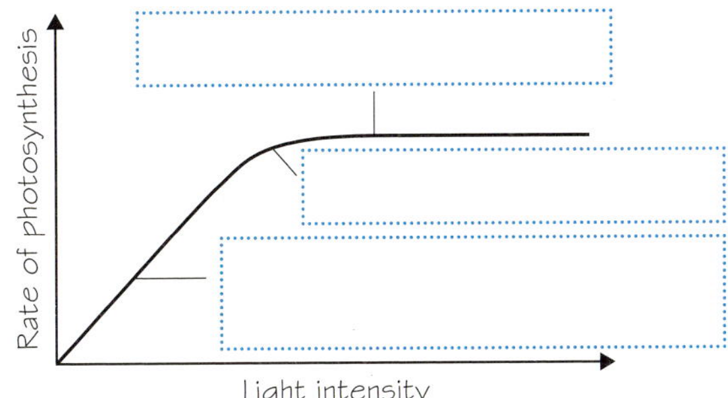

Other factors, such as carbon dioxide levels, begin to limit the reaction.

As light intensity increases, the rate of photosynthesis increases.

No further increase can occur without more carbon dioxide.

2 The graph on the right shows how light intensity affects the rate of photosynthesis at different carbon dioxide concentrations.

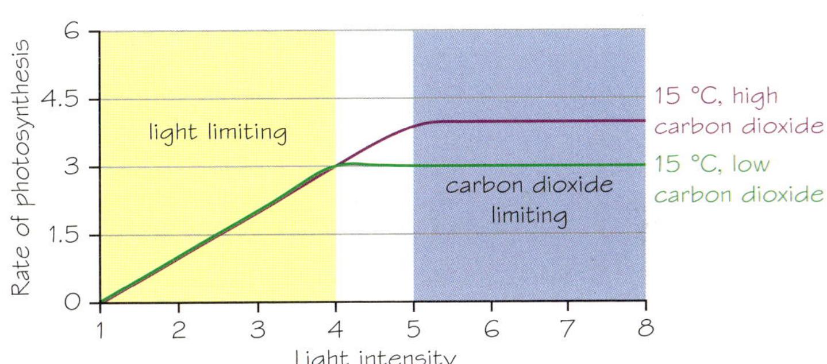

Complete the table by circling the correct words shown in bold.

Area of graph	Description	Likely limiting factor
Yellow	As the light intensity increases the rate of photosynthesis **increases / stays the same**. As the concentration of carbon dioxide increases the rate of photosynthesis **increases / stays the same**.	**Light / carbon dioxide** must be the limiting factor because the graph shows that as light intensity increases, the rate of photosynthesis increases.
Blue	As the light intensity increases the rate of photosynthesis **increases / stays the same**. As the carbon dioxide increases the rate of photosynthesis **increases / stays the same.**	**Light / carbon dioxide** must be the limiting factor because the rate of photosynthesis is higher for the high carbon dioxide concentration at the same light intensity.

3 Draw a line on the graph in question **2** to show a predicted rate of photosynthesis at an even higher carbon dioxide concentration.

Sample response

You could be asked to explain how different tissues in a plant are adapted to their function. Make sure you are confident explaining adaptations of the different types of tissue in a plant and how water is transported.

You also need to be able to explain graphs showing how factors can interact to affect the rate of photosynthesis.

Look at this exam-style question and student response.

Exam-style question

1 Figure 1 shows the cross-section of a leaf.

(a) The upper epidermis of a leaf is thin and transparent.

Explain how this helps the plant to survive. **(2 marks)**

(b) Explain **one** way in which the structure of the layer of palisade cells is adapted to carry out its job. **(2 marks)**

(c) Explain how transpiration occurs. **(3 marks)**

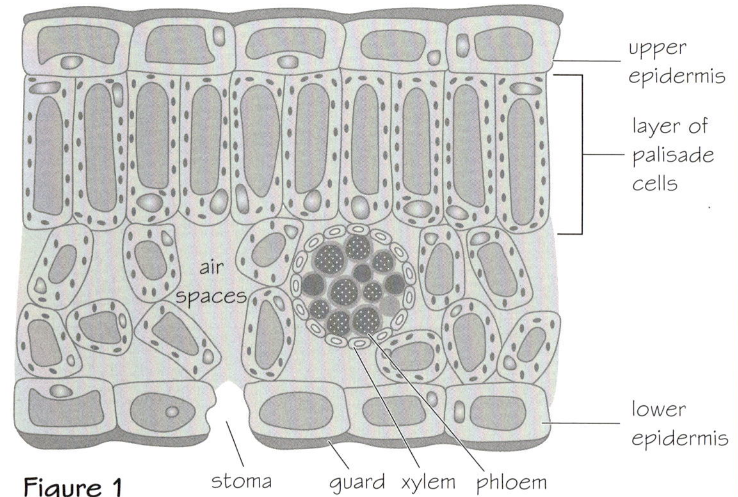

Figure 1

(a) *Being thin and transparent allows more light to reach the layer of palisade cells.*

(b) *They contain lots of chloroplasts.*

(c) *Water evaporates and moves out of the leaf through openings in the leaf.*

This causes more water to be drawn up inside the hollow tube inside the stem.

As water moves up the xylem it causes more water to enter the roots by active transport.

1 The student's answer to part **(a)** is worth only 1 mark. What needs to be included in their answer to get the second mark?

2 What important detail has the student missed in their answer to part **(b)**?

Remember to use **physical feature + linking word + explanation** when you explain how a feature is adapted to its function.

3 Improve the answer to part **(c)** by replacing the parts highlighted in green with your own answers.

Your turn!

It is now time to use what you have learned to answer the exam-style question from page 9.

Remember to read the question thoroughly, looking for information that may help you.

Make good use of your knowledge from other areas of biology.

Exam-style question

1 The lower epidermis of a leaf from a bay tree contains many guard cells.

(a) Explain how guard cells help the bay tree survive. **(2 marks)**

....

....

....

....

Guard cells form small holes called stomata. Guard cells can open and close to control the passage of water. When the plant is short of water the guard cells become shrunken, closing the stoma.

You don't need to know about bay trees to answer this question. The question is testing your knowledge and understanding of plant structure and function.

Figure 1 shows how carbon dioxide concentration affects the rate of photosynthesis in a bay tree.

(b) What is the maximum rate of photosynthesis of the bay tree shown by the graph? **(1 mark)**

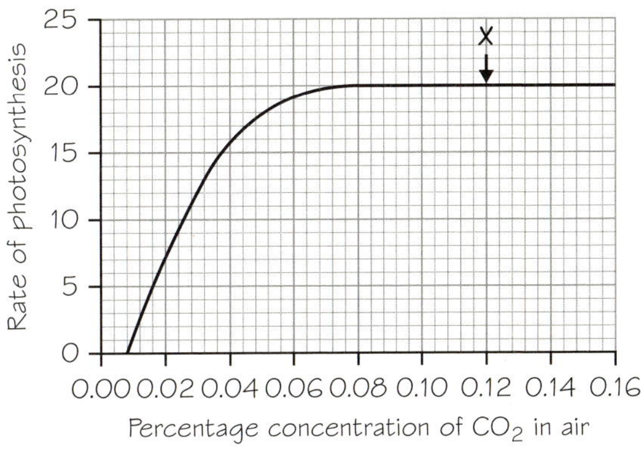

....

....

Figure 1

Before you attempt to answer a question like this, spend time making sense of the graph. When the line levels out, it means a limiting factor has stopped the rate of photosynthesis increasing.

(c) Suggest **one** factor that could be limiting the rate of photosynthesis at point X. **(1 mark)**

....

Remember The mnemonic for the factors affecting the rate of photosynthesis is: Leprechauns Cook Toast (Light, Carbon dioxide, Temperature).

Need more practice?

Exam questions may ask about different parts of a topic, or parts of more than one topic. Questions about plant biology could occur as:

- questions about plant cells, tissues or organs
- questions about an investigation into leaf structure, transpiration or photosynthesis.

Have a go at this exam-style question. 

Exam-style question

1 Plants contain phloem tissue.

Figure 1 shows a cross-section of some phloem cells.

Phloem transports sucrose solution around a plant.

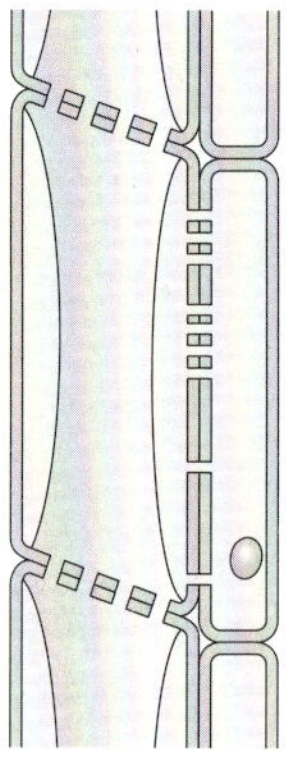

(a) Explain **two** ways phloem tissue is adapted to its function. **(2 marks)**

Figure 1

(b) Xylem tissue plays an important role in transpiration.

What is meant by transpiration? **(2 marks)**

Boost your grade

You may be expected to apply your understanding to an unfamiliar situation so make sure you know the plant biology facts really well. Often, plant biology questions are linked to plants you might not have heard of or to farming, so make sure you can confidently explain the effects of different factors on the rate of photosynthesis.

How confident do you feel about each of these **skills**? Colour in the bars.

1 How do I explain how leaf structures are adapted to their functions?

2 How do I explain how water is transported in plants?

3 How do I explain how limiting factors interact?

Get started AO1, AO2

3 Genetics

This unit will help you to use genetic terms correctly and construct Punnett squares to make predictions.

In the exam, you will be asked to answer questions such as the one below.

Exam-style question

1 Fruit flies can have normal or vestigial wings.

A vestigial wing is an abnormally small wing.

Normal-winged fruit flies have the genotype **NN** or **Nn**.

Vestigial-winged flies have the genotype **nn**.

(a) What causes vestigial wings?

Tick **one** box.

a homozygous dominant genotype ☐

a homozygous recessive genotype ☐

a heterozygous genotype ☐

the environment ☐ **(1 mark)**

A male and a female normal-winged fruit fly were crossed.

They produced 52 offspring. 39 had normal wings and 13 had vestigial wings.

(b) Draw a Punnett square to show how the ratio of normal-winged and vestigial-winged offspring was produced. **(2 marks)**

....

(c) Determine the probability of two heterozygous normal-winged flies producing a gamete with the genotype **nn**. **(1 mark)**

....

Biology

You will already have done some work on genetics. Before starting the **skills boosts**, rate your confidence for each skill. Colour in the bars.

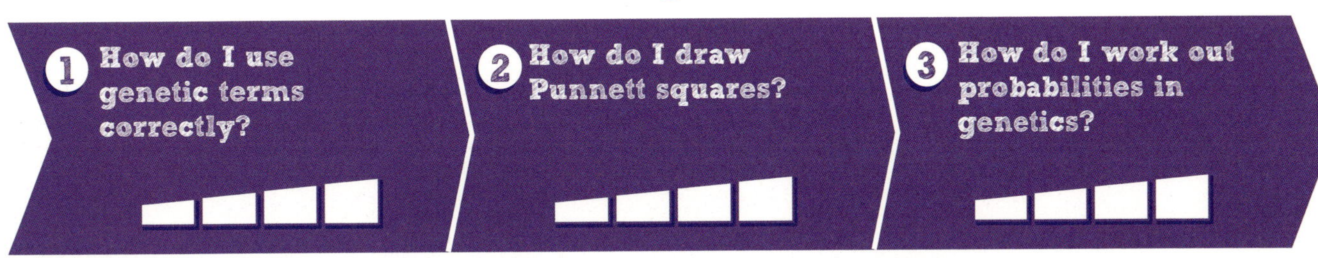

1 How do I use genetic terms correctly?
2 How do I draw Punnett squares?
3 How do I work out probabilities in genetics?

Get started

During fertilisation, one set of chromosomes comes from the female and one set from the male.

Body cells have two versions of each chromosome, arranged in pairs. Chromosomes contain genes that control the characteristics of an organism.

Different genes control different characteristics such as hair or eye colour. The genes on a pair of chromosomes may be different because we inherit these genes from both parents.

Different versions of the same gene are called **alleles.** If both alleles for one gene are the same, the organism is **homozygous** for the characteristic. If the alleles are different, the organism is **heterozygous**.

1 Draw a line to link each key word with the correct definition.

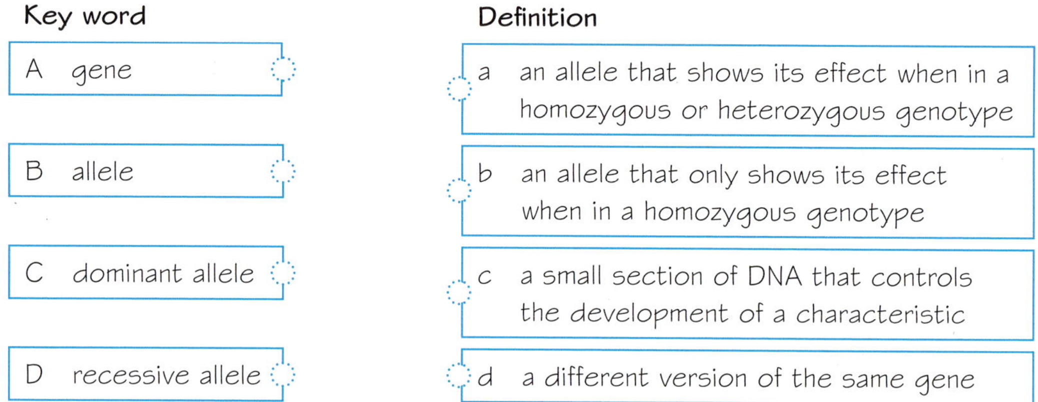

Key word
A gene
B allele
C dominant allele
D recessive allele
Definition
a an allele that shows its effect when in a homozygous or heterozygous genotype
b an allele that only shows its effect when in a homozygous genotype
c a small section of DNA that controls the development of a characteristic
d a different version of the same gene

The alleles in an organism are its **genotype**. The physical appearance resulting from the alleles is its **phenotype**.

Scientists use letters to represent different alleles. A dominant allele is shown using a capital letter and a recessive allele is shown using a lower-case letter. The dominant allele is always written before the recessive allele. For example **Aa,** not **aA.**

2 Tongue-rolling in humans is controlled by a dominant allele (**T**). Non-rolling is controlled by the recessive allele (**t**). A student is **homozygous** for non-tongue-rolling.

a Write the student's **genotype**. ....

b Write the student's **phenotype**. ....

A student is asked to write the genotype of a heterozygous tongue-roller. They write tT.

c Write the correct genotype. ....

> **Remember** A single gene controls tongue-rolling, but most characteristics are a result of several genes interacting.

Punnett squares show the possible combinations of genotypes that can occur in offspring when two organisms are crossed. These diagrams can be used to predict the probability of different phenotypes. The diagram below shows the possible genotypes of the offspring produced by a brown-eyed parent and a blue-eyed parent.

3 Identify the following in the diagram.

a Circle a homozygous dominant individual.

b Underline a homozygous recessive individual.

c Highlight a heterozygous individual.

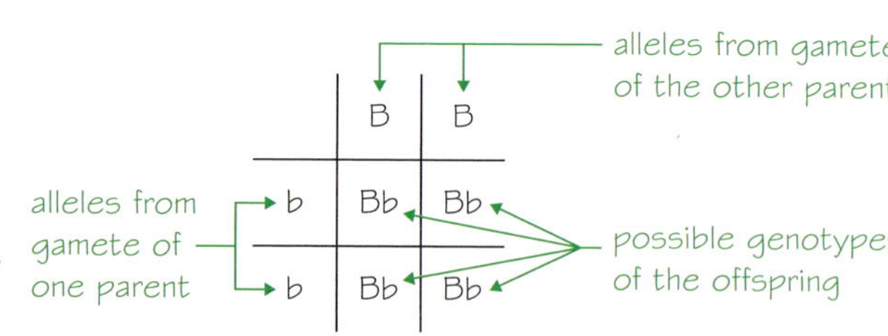

How do I use genetic terms correctly?

There are many genetic terms you need to know, understand and be able to use correctly. Many of the terms in genetics are closely linked so it is important to make sure you understand each term and how they are linked.

1 The diagram shows a pair of chromosomes found in a body cell. Add the following labels to the diagram.

different genes heterozygous allele pair (Dd) homozygous recessive allele pair alleles of gene A homozygous dominant allele pair

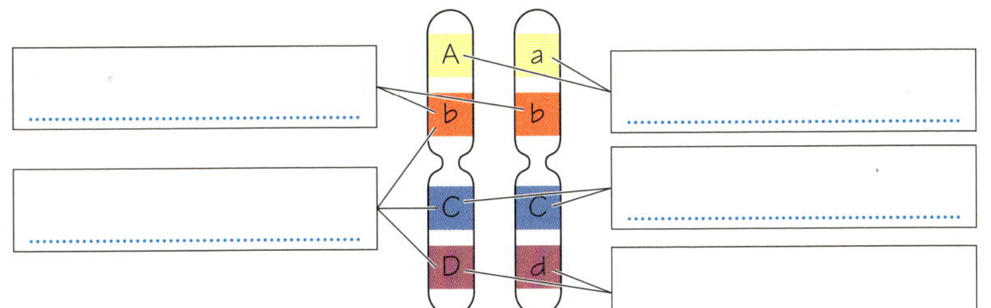

Remember 'homo' means the same and 'hetero' means different.

Chromosomes are structures made from very long and tightly coiled molecules of DNA.

2 Explain why chromosomes are found in pairs in body cells. Use the words **fertilisation** and **gamete** in your answer.

....

....

3 A rabbit receives the allele for white fur from its mother and the allele for black fur from its father. Black is the dominant allele.

a What is the genotype of the rabbit? ....

It does not matter what letter is chosen to represent an allele. Often, the first letter of the dominant allele is used, but this is not always the case.

Remember The dominant allele is always written with a capital letter. The letter representing the recessive allele should be the same letter in lower case.

b What is the phenotype of the rabbit? ....

genotype tells us about the **g**enes, **p**henotype tells us the **p**hysical characteristics.

4 Use the sentence starters to explain what determines whether a rabbit has white or black fur.

A rabbit inherits one chromosome from its mother and ....

The chromosomes contain a gene that ....

The gene comes in different versions called ....

Black is a dominant allele for fur colour, so if the rabbit inherits the allele for black fur from its mother or father ....

The allele for white fur is recessive so ....

Biology

How do I draw Punnett squares?

Punnett squares help us to understand the possible outcomes when organisms produce offspring. You need to be able to use Punnett squares to predict and explain the outcomes of single gene crosses between organisms.

The allele for tall pea plants is **T**. The allele for short pea plants is **t**. Two pea plants are crossed. One plant is homozygous dominant for being tall. The other plant is homozygous recessive for being short.

1 Complete the diagram to show the genotypes of the plants and their gametes.

Parent plant phenotypes	Tall	Short
Parent plant genotypes	TT	....
Gamete genotypes	....	t t

Remember Gametes are produced by meiosis so they only contain **one** allele, not two, for each gene. When the gametes fuse in fertilisation the offspring get one allele for each gene from each parent.

2 Write the genotype from one parent plant along the top of the Punnett square and the genotype of the other parent plant down the left-hand side.

When you write the gamete genotypes along the top and the side of the Punnett square make sure you separate the letters. You should only write one allele on the side of each box.

It does not matter which parent is on the side or the top of the Punnett square.

3 Complete the Punnett square to show the possible alleles for the offspring.

Fill in the letters from the top and side that line up with the square.

Remember A dominant allele is always written first. For example, Tt, not tT.

4 What is the **phenotype** of the offspring? Tick the correct answer.

Short ☐ Tall ☐ Medium ☐ Tt ☐

Remember Phenotype is the physical appearance that results from the alleles.

5 Write a sentence explaining your answer using the following key words.

dominant	recessive	overrules	heterozygous

....

....

6 On the Punnett square in question 2:

a Circle the **genotypes** of the offspring that will have a tall **phenotype**.

b Explain why a pea plant would be short.

....

....

 How do I work out probabilities in genetics?

The outcomes of genetic crosses are given as ratios. These ratios can be used to work out the theoretical probability of having offspring with certain characteristics.

1. A plant has two alleles for flower colour. The allele for red flowers (**R**) is dominant to the allele for white flowers (**r**). The Punnett square shows a cross between two heterozygous plants.

	R	r
R	RR	Rr
r	Rr	rr

   a How many possible outcomes are shown in the diagram?

    ....

   b How many outcomes will produce plants with white flowers?  ....

   c Give your answer to part **b** as:

   i a fraction of the possible outcomes ....

   ii a percentage of the possible outcomes ....

   iii a decimal. ....

2. Complete the table to show the probabilities of the outcomes producing a plant with red flowers.

Phenotype	Probability		
	Fraction	Decimal	Percentage
red flowers		0.75	

Probability is a measure of how likely something is to happen. Probabilities can be written as fractions, decimals or percentages.

$$\text{Probability of an event happening} = \frac{\text{number of ways it can happen}}{\text{total number of outcomes}}$$

To convert a fraction into a decimal you divide the top number by the bottom number (divide the numerator by the denominator).

A probability expressed as a decimal can be converted into a percentage by multiplying the decimal by 100.

Once you know the ratio you can easily convert it into a probability. For example, if there is a ratio of white flowers to red flowers of 1 : 3, that is equivalent to 1 in 4 or $\frac{1}{4}$.

3. Guinea pigs can have rosetted or smooth fur. The allele for rosetted fur is dominant (**R**) to smooth fur (**r**). A homozygous recessive guinea pig (rr) is bred with a heterozygous guinea pig (Rr).

   a Fill in the Punnett square to show the possible outcomes.

Probability is always between 0 and 1. An event that is impossible has a probability of 0. An event that is certain to happen has a probability of 1 (or 100%).

   b On your diagram, circle the genotype of the offspring that have **rosetted** fur.

   c What is the ratio of rosetted fur to smooth fur? ....

   d Convert the ratio into a probability. Express the probability as:

   i a fraction ....

   ii a decimal ....

   iii a percentage. ....

Biology

Sample response

Your understanding of genetic inheritance will often be tested using examples of inherited characteristics in plants and animals. Some of these examples may use characteristics you are unfamiliar with. Make sure you are confident using a Punnett square to make predictions for a variety of examples.

Look at this exam-style question and the answers given by a student.

Exam-style question

1 Eye colour is controlled by alleles of several genes.

The dominant allele of the gene (**B**) produces brown eyes and the recessive allele (**b**) produces green eyes.

A man and a woman are both heterozygous for brown eyes. They have a child with green eyes.

(a) Complete the Punnett square to show why their child came to have green eyes. **(3 marks)**

(b) Calculate the probability of the child having a phenotype of green eyes. **(1 mark)**

(c) Explain the difference between a genotype and a phenotype. **(2 marks)**

(a)

	B	g
B	BB	Bg
g	Bg	gg

(b) *1:3 so 1/3 or 33%*

(c) *A genotype is the type of alleles in an organism and the phenotype means colour of the eyes.*

1 **a** A capital letter 'B' has been used to represent the dominant allele. The student has used a 'g' to represent the recessive allele for green eyes. What letter should the student have used for the recessive allele? ....

The student has **not** calculated the correct probability. They have not converted the ratio into the correct probability.

b Convert the green : brown ratio of 1 : 3 into a fraction. ....

c What is the probability of the phenotype of green eyes expressed as a:

i decimal ....

ii percentage. ....

d The student gains only 1 mark out of 2 for his answer to part (c).

Highlight the parts of the answer that are correct and underline the part of the answer that needs to be changed to gain the second mark.

e Write what the student needs to say to gain the second mark.

....

....

Has the student answered the question? Are the meanings of both terms 'genotype' and 'phenotype' made clear? Is eye colour an example of a phenotype or a definition?

Your turn!

It is now time to use what you have learned to answer the exam-style question from page 17.

Remember to read the question thoroughly, looking for information that may help you.

Make good use of your knowledge from other areas of biology.

Exam-style question

1 Fruit flies can have normal or vestigial wings.

A vestigial wing is an abnormally small wing.

Normal-winged fruit flies have the genotype **NN** or **Nn**.

Vestigial-winged flies have the genotype **nn**.

(a) What causes vestigial wings?

Tick **one** box.

a homozygous dominant genotype ☐

a homozygous recessive genotype ☐

a heterozygous genotype ☐

the environment ☐

Use the information given about the genotypes of the fruit fly to help you answer this question.

(1 mark)

Genetic questions might use organisms or characteristics you have not heard of. This question is testing your ability to apply your knowledge and understanding of genetics. You don't actually need any knowledge about fruit flies or vestigial wings.

A male and a female normal-winged fruit flies were crossed.

They produced 52 offspring. 39 had normal wings and 13 had vestigial wings.

(b) Draw a Punnett square to show how the ratio of normal-winged and vestigial-winged offspring were produced.

(2 marks)

Before you draw the diagram, look at the numbers of flies that had normal wings versus those that had vestigial wings. Is there a ratio? Can the ratio be simplified? A ratio of 39 : 13 can be simplified to 3 : 1.

A cross between two parents with a heterozygous genotype is likely to produce a 3 : 1 ratio.

(c) Determine the probability of two heterozygous normal-winged flies producing a gamete with the genotype **nn**. **(1 mark)**

....

Remember Once you have calculated the ratio, it is easy to convert into a probability. The probability can be expressed as a fraction, a decimal or a percentage.

Biology

Need more practice?

Exam questions may ask about different parts of a topic, or parts of more than one topic. Questions about genetics could occur as:

- questions about physical characteristics
- questions about sex determination or inherited conditions.

Have a go at this exam-style question. 

Exam-style question

1 Polydactyly is an inherited condition in which a person has extra fingers or toes.

Polydactyly is caused by a dominant allele **(D)**.

(a) Which of the following statements best describes an **allele**?

Tick **one** box.

a mutant gene	☐	a type of gamete	☐
a long chromosome	☐	a variant of a gene	☐

(1 mark)

Remember Genes exist in different forms, for example, brown fur or white fur.

Tom has polydactyly. He has a **heterozygous** genotype. His wife, Beth, does not have polydactyly.

(b) State Tom's genotype? .... **(1 mark)**

(c) State Beth's genotype? .... **(1 mark)**

Use the information given in the question to work out the genotype. Tom has polydactyly and he is heterozygous. Beth does not have polydactyly so she must be homozygous recessive.

(d) Draw a Punnett square to calculate the probability of Tom and Beth having a child with polydactyly. Express the probability as a percentage.

Probability of offspring having polydactyly:

....

(4 marks)

Show the possible genotypes of the children in a Punnett square. Of the four possible gametes, how many would have polydactyly?

Boost your grade

To improve your grade, make sure you can:

- confidently explain the differences between different genetic terms and how they are linked
- interpret family trees to work out genotypes, phenotypes and probabilities of offspring having a particular genotype.

How confident do you feel about each of these **skills**? Colour in the bars.

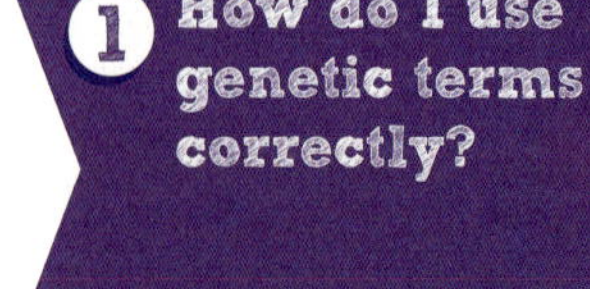

1 How do I use genetic terms correctly?

2 How do I draw a Punnett square?

3 How do I work out probabilities in genetics?

Get started AO2, AO3

Hormones

This unit will help you to explain how the hormones of the menstrual cycle interact. It will also help you to understand how negative feedback works and how hormones are used to help couples have children during a process called IVF.

In the exam, you will be asked to answer questions such as the one below.

Exam-style question

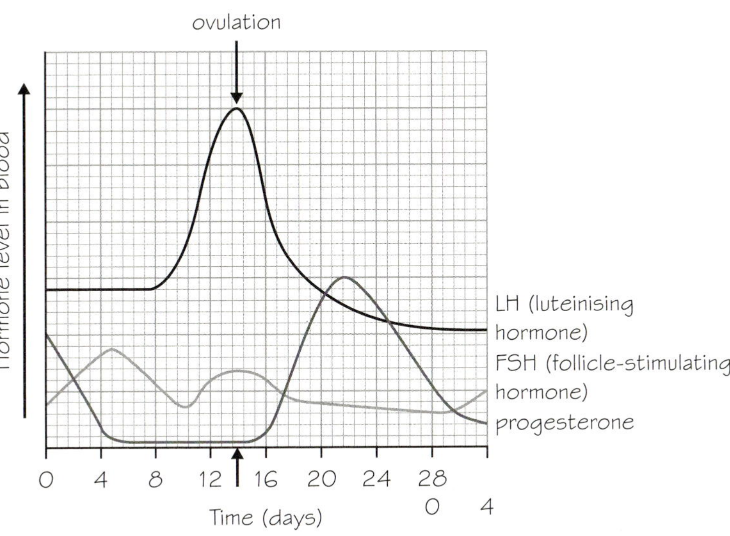

1 (a) Suggest the stimuli that might cause an egg to be released. Use information from the graph to justify your answer. **(3 marks)**

....

(b) A high level of oestrogen can inhibit the production of FSH by the pituitary gland.

Explain how this is an example of negative feedback. **(2 marks)**

....

(c) A drug called clomifene is used to block the inhibitory effect of oestrogen on FSH production.

Explain how this may help in the treatment of infertility. **(2 marks)**

....

You will already have done some work on hormones. Before starting the **skills boosts**, rate your confidence in explaining the menstrual cycle, negative feedback and IVF. Colour in the bars.

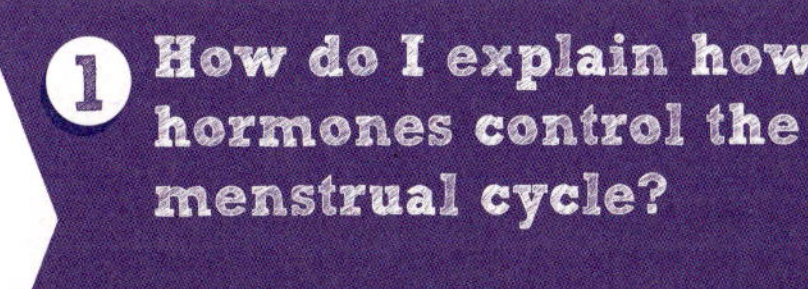

1 How do I explain how hormones control the menstrual cycle?

2 How do I explain negative feedback?

3 How do I explain the use of hormones in IVF?

Biology

Get started

Hormones are chemical transmitters that are involved in the control of many bodily processes including the menstrual cycle. They regulate these processes using a form of control called **negative feedback**. Some hormones can be used to bring about changes in a woman's reproductive system and enable successful pregnancy.

1 Complete the sentences below about hormones. Select the correct words from the box.

blood	respiration	skin	hormones	glands	organs	ovulation	oestrogen

Many bodily processes are controlled by .... These chemicals are released by .... and are transported in the .... to their target .... One example is ...., which is released by the ovaries and brings about ....

Negative feedback is a form of hormonal control that works a bit like the thermostat in your house. If the house gets too warm, the thermostat turns the heating off. The house cools down. When the house gets too cold, the thermostat turns the heating on again, and so on.

2 Complete the flow diagram illustrating negative feedback by writing the correct numbers in the boxes.

Blood glucose is also controlled by negative feedback. If blood glucose increases above normal, cells in the pancreas release insulin, which stimulates the liver to remove glucose from the blood. If it decreases below normal, other cells in the pancreas release glucagon, which stimulates the liver to release glucose into the blood.

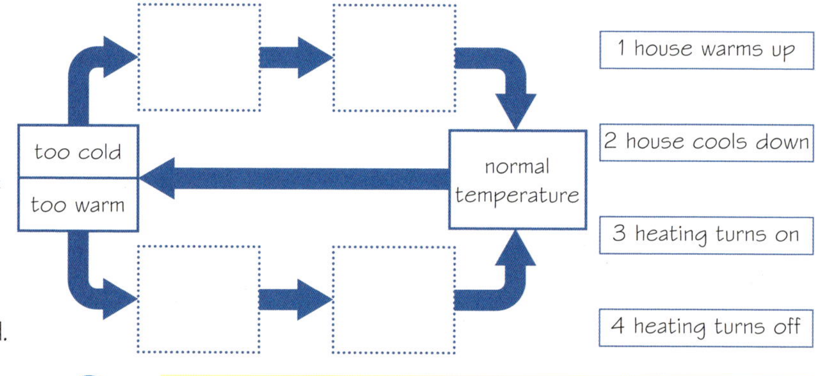

3 Complete the flow diagram by writing the **four** labels in the correct boxes.

Stimulate means 'to start or increase production of'. Inhibit means 'to block or decrease production of'.

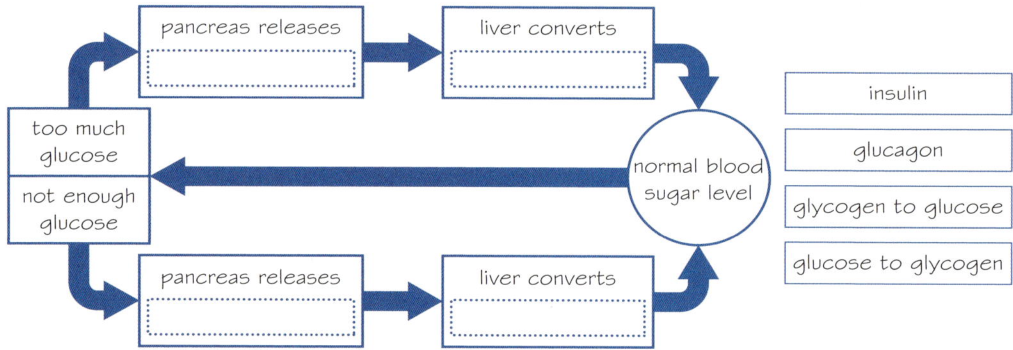

Some couples are unable to have a child naturally and this can be due to irregular levels of reproductive hormones. The negative feedback can be too great so that levels of some hormones are inhibited. This can be helped by Assisted Reproductive Technology (ART). One technique is ***in-vitro* fertilisation (IVF)**. IVF treatment helps infertile women to become pregnant.

4 Circle the correct bold words in the sentences about IVF.

The eggs are collected from the mother's **ovary / uterus**. Each egg is fertilised by a sperm. Each fertilised egg develops into a ball of cells called an embryo. One or two of these embryos are inserted into the mother's **ovary / uterus**.

How do I explain how hormones control the menstrual cycle?

The menstrual cycle is a cycle of changes in a woman's reproductive system over about 28 days. Menstruation occurs between days 1 and 5 and then four hormones (**FSH**, **LH**, **oestrogen** and **progesterone**) interact to control the maturation and release of a new egg cell from the ovary.

1. Write the **four** hormones in the correct boxes in the table below. Use the graph to help you.

FSH LH oestrogen progesterone

....	Increases slightly around day 10–11 before the egg starts to mature in the follicle.
....	Increases after ovulation on day 14.
....	Peaks sharply as the egg is released and then continues to decrease.
....	Starts to decrease as progesterone levels decrease.

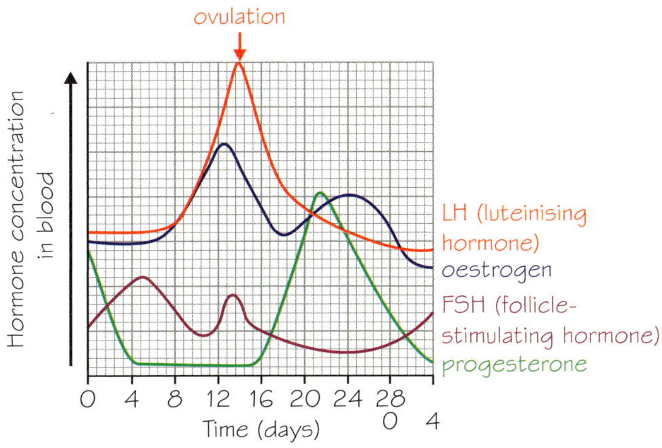

2. Using information from the diagram below, number the statements in the correct order (1 to 10) to outline the process of the menstrual cycle. The first one has been done for you.

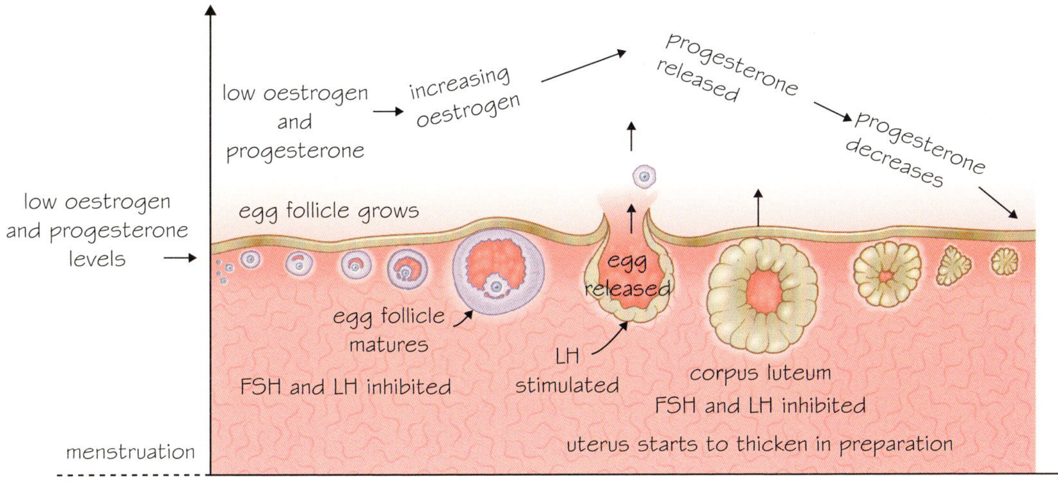

The egg follicle is the part of the ovary where an egg cell grows. Corpus luteum is formed from the mature egg follicle once the egg has been released. It then starts to release progesterone.

Maturing follicle stimulates oestrogen production	☐	Increased levels of oestrogen stimulate LH production	☐
Follicle forms corpus luteum and releases progesterone	☐	Increase in LH stimulates egg release (ovulation)	☐
Oestrogen and progesterone levels start to decrease	☐	Increasing oestrogen stimulates thickening of uterus lining	☐
FSH and LH release from pituitary is inhibited	☐	Pituitary releases FSH and LH	☐
Low oestrogen and progesterone stimulate menstruation	1	FSH stimulates growth and maturation of egg follicle	☐

How do I explain negative feedback?

Negative feedback occurs when:

- an **increase** in the concentration of one hormone directly causes changes that result in a **decrease** in the concentration of another hormone
- a **decrease** in the concentration of one hormone directly causes changes that result in an **increase** in the concentration of another hormone.

Thyroxine is released by the thyroid gland. It helps to regulate the rate at which energy is transferred from your food to chemical reactions in many types of cells. Thyroxine production is regulated by negative feedback, so that its increase results in the decrease of other hormones, and its decrease results in the increase of other hormones.

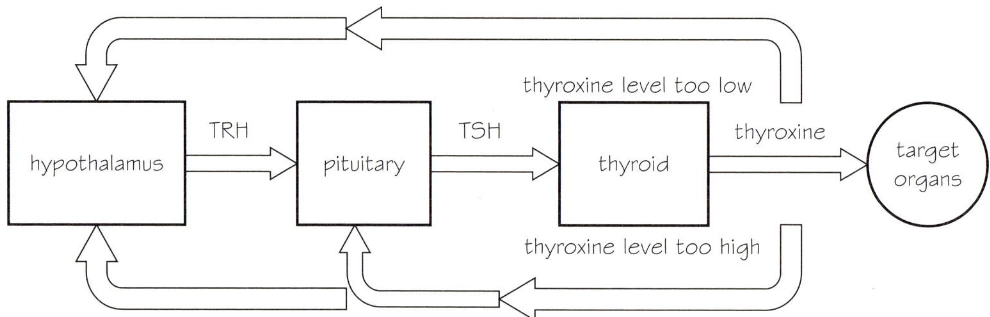

1 **a** Draw linking lines from the glands to the hormones they release. Use the flow chart to help you.

Thyrotropin-releasing hormone (TRH) is released when the hypothalamus is stimulated by low levels of thyroxine.

Gland	Hormone released
A thyroid	a thyrotropin-releasing hormone (TRH)
B hypothalamus	b thyroid-stimulating hormone (TSH)
C pituitary	c thyroxine that helps regulate metabolic rate

TRH stimulates the pituitary gland to release thyroid-stimulating hormone (TSH).

b On the flow chart above:

i Colour the glands in yellow. 

ii Circle the hormones in blue. 

iii Colour the arrows that show stimulation of thyroxine production in green. 

iv Colour the arrows that show inhibiting of thyroxine production in red. 

2 Write down whether these statements about negative feedback are true or false. 

a If the thyroxine level is lower than normal in the blood, then it stimulates the hypothalamus to release thyrotropin-releasing hormone (TRH). ....

b TRH released from the hypothalamus stimulates the release of thyroid-stimulating hormone (TSH) by the pituitary gland. ....

c TSH produced by the pituitary gland will cause a decrease in thyroxine production by the thyroid gland. ....

d If the thyroxine level is higher than normal in the blood, then glands in the brain called the hypothalamus and pituitary are stimulated. ....

Use the flow chart and follow the arrows.

3 How do I explain the use of hormones in IVF?

Fertilisation requires both egg and sperm. Sometimes, there is a problem with the production of eggs. This can be caused by hormones. By correcting the hormone balance, egg production can be re-established and successful IVF treatment can take place.

1 Clomifene is a drug that can be used to treat infertility. Look at the graph below.

Look at how oestrogen, FSH and LH levels interact on the graph.

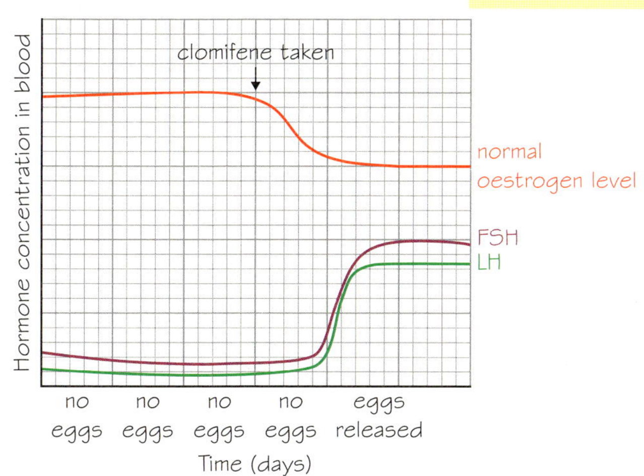

a Describe the effect that clomifene has on:

i normal oestrogen level. ....

ii FSH and LH concentration. ....

b Describe how the initial oestrogen level affects the FSH and LH levels.

....

c Explain why eggs start to be released after clomifene is taken.

....

What does clomifene do to the LH and FSH levels?

2 Write numbers (1 to 6) in the boxes to show the correct order in which clomifene affects egg production. The first stage has been done for you.

Egg(s) released	☐	Normal level of oestrogen inhibited	☐
Pituitary gland releases FSH and LH	☐	FSH and LH stimulate ovary	☐
Ovary starts to ripen and mature eggs	☐	50 mg clomifene tablet taken each day	1

3 A woman takes clomifene but after several months it has had no effect.

a Name the hormone that clomifene affects. ....

Look at the graph above.

b If the clomifene is working properly, state which hormones are not working properly.

....

c Explain why this might be the case.

Which gland produces these hormones?

....

4 The diagram illustrates the stages in IVF treatment. Explain how this diagram shows what scientists do to bring about a successful fertilisation.

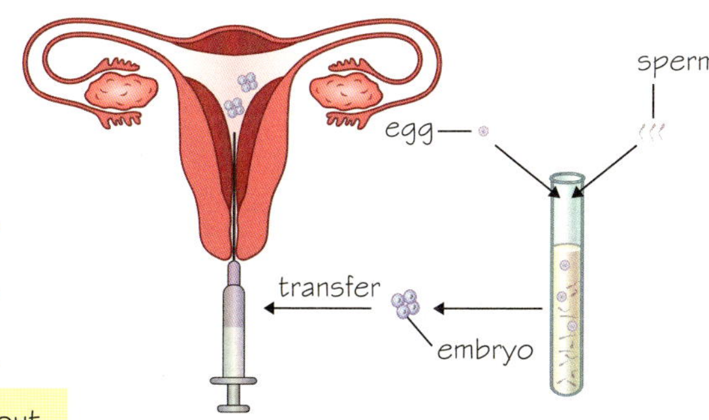

....

....

....

Think about the order in which the stages will be carried out.

Sample response

Look at this exam-style question and use the student response to improve your understanding. Consider what the graph is showing and use your knowledge to try to identify the hormones involved. Think about the interaction between the hormones, whether negative feedback is being used to control their actions and how might this help.

Look at this exam-style question and student response.

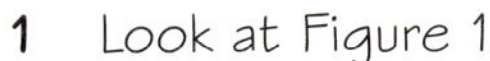

Exam-style question

1 Look at Figure 1.

(a) Name hormones **W** and **X**. **(2 marks)**

W: LH

X: progesterone

(b) Explain how control of LH is an example of negative feedback. **(4 marks)**

LH causes ovulation and a corpus luteum to form. The corpus luteum releases progesterone.

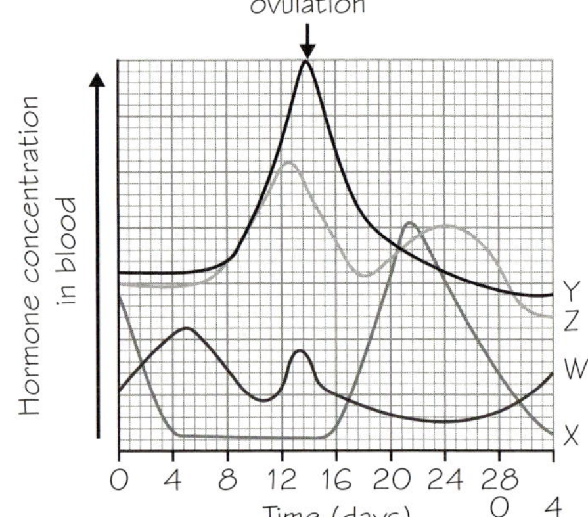

Figure 1

(c) *In-vitro* fertilisation (IVF) treatment uses some of the hormones shown in Figure 1. Explain why using FSH and LH increase the likelihood of some women becoming pregnant. **(2 marks)**

FSH causes eggs to mature and LH causes the eggs to be released.

1 a One of the answers to part (a) is wrong. Circle (A) the wrong hormone.

b Write down what the correct hormone should be.

....

2 The student only got 2 marks for part (b). To get full marks, they needed to explain the role of progesterone and LH.

a Describe what increased progesterone does.

....

....

b Explain how LH causes a reduction in progesterone.

....

....

3 In their response to part (c), the student described what FSH and LH do. Explain why using normal levels of FSH and LH will increase chances of pregnancy.

....

....

Your turn!

It is now time to use what you have learned to answer the exam-style question from page 25. Remember to read the question thoroughly, looking for information that may help you. Make good use of your knowledge from other areas of biology.

Exam-style question

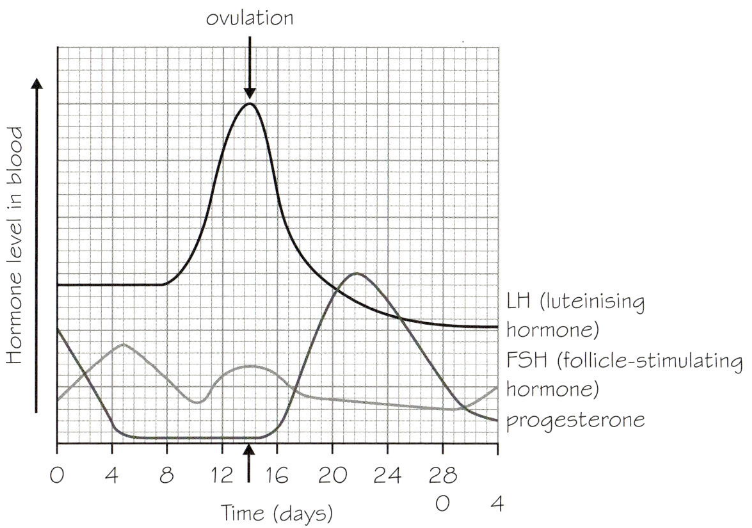

1 **(a)** Suggest the stimuli that might cause an egg to be released. Use information from the graph to justify your answer.

Which two hormones increase immediately before ovulation?

(3 marks)

....

....

(b) A high level of oestrogen can inhibit the production of FSH by the pituitary gland.

Explain how this is an example of negative feedback.

What will happen to FSH levels now?

(2 marks)

....

....

(c) A drug called clomifene is used to block the inhibitory effect of oestrogen on FSH production.

Explain how this may help in the treatment of infertility.

How will the correct FSH levels now affect fertility?

(2 marks)

....

....

Need more practice?

Exam questions may ask about different parts of a topic, or parts of more than one topic. Questions about hormones could occur as:

- questions on how a treatment or drug might influence the hormones
- questions about an experiment or investigation.

Have a go at this exam-style question.

Exam-style question

1 A couple are trying for a baby without success. Doctors monitor the woman's progesterone, thyroxine and follicle-stimulating hormone (FSH) levels.

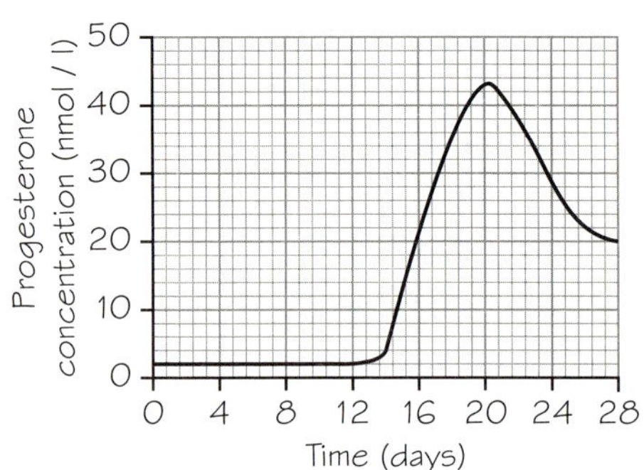

(a) The graph shows the woman's progesterone levels over 28 days. The level changed following day 14. Explain why this happened. **(2 marks)**

....

....

(b) Using information from the graph, explain whether the woman is pregnant. **(2 marks)**

....

....

(c) Doctors notice that the woman's pituitary gland does not release enough FSH. Explain how injections of FSH could increase her chances of having a baby. **(3 marks)**

....

....

(d) The woman has an underactive thyroid gland. Explain how this could affect her metabolic rate. **(3 marks)**

....

....

Boost your grade

Practise drawing sketch graphs of menstrual cycle hormone concentrations over 28 days.
Draw the negative feedback flow diagram next to an exam question before answering.
Be able to explain how an IVF drug such as clomifene affects the menstrual hormones.

How confident do you feel about each of these **skills**? Colour in the bars.

1 How do I explain how hormones control the menstrual cycle?

2 How do I explain negative feedback?

3 How do I explain the use of hormones in IVF?

5 Genetic modification

This unit will help you to understand the impacts of selective breeding and describe the stages of genetic engineering. It will also help you to explain the potential benefits and risks of genetic engineering.

In the exam, you will be asked to answer questions such as the one below.

Exam-style question

1 Scientists have used selective breeding to produce a new variety of wheat plant that is better quality and produces a higher yield than wild wheat plants.

The new variety of wheat plant has fewer different alleles than wild wheat plants.

(a) Explain why the process of selective breeding causes a reduction in the number of different alleles in the population of wheat plants. **(2 marks)**

....

(b) Give **one** reason why it can be a disadvantage for a variety of wheat to have fewer different alleles. **(1 mark)**

....

Wheat plants can be genetically engineered. This type of wheat is called genetically modified (GM) wheat. Some types of GM wheat are resistant to herbicides.

(c) Give **two** reasons why a farmer might benefit from growing GM wheat. **(2 marks)**

....

(d) Some people are concerned about the use of GM wheat. Suggest **two** reasons why. **(2 marks)**

....

You will already have done some work on genetic modification. Before starting the **skills boosts**, rate your confidence in each area. Colour in the bars.

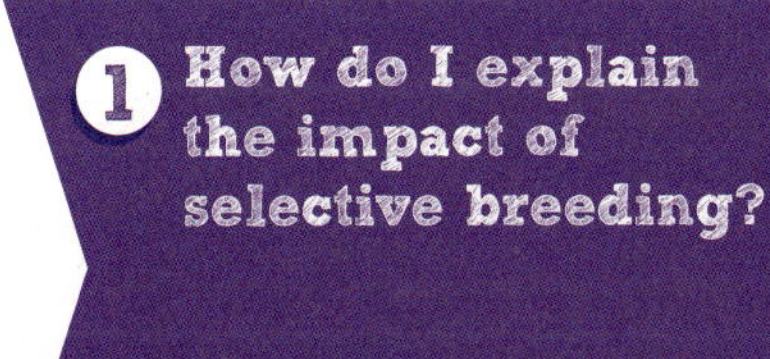

1 **How do I explain the impact of selective breeding?**

2 **How do I describe the stages of genetic engineering?**

3 **How do I explain the benefits and risks of selective breeding and genetic engineering?**

Get started

Selective breeding (artificial selection) is the process by which humans breed plants and animals for particular genetic characteristics.

This is done by selecting parents with the desired characteristics and breeding them together. From the offspring produced, only those with the desired characteristic are bred together. This process is repeated until all the offspring show the desired characteristic.

1 Draw lines to link the key words with their correct meanings.

Key word	Meaning
A species	a a group of one species living in the same area
B breed	b a group of plants of the same species that have characteristics that make them different from other members of the species
C variety	c a group of animals of the same species that have characteristics that make them different from other members of the species
D population	d a group of organisms that can reproduce with each other to produce fertile offspring

2 Number the statements in the table to show the correct order of the stages in the selective breeding of chickens. The first one has been done for you.

Description of stage in selective breeding	Stage
Chickens are bred over many generations until all of the chicks produced have the desired characteristic	
Chickens with the desired characteristic are chosen	1
Offspring with the desired characteristic are bred together	
Parents with the desired characteristic are bred together	

3 Genetic engineering is the process of changing the DNA of one organism (its **genome**) by inserting genes from another. It creates a **genetically engineered organism**.
What is the goal of genetic engineering?
Tick **one** box.

A To create a new species

B To add new desirable characteristics

C To add genes that stop DNA mutating

D To isolate and remove defective alleles in an organism

4 Enzymes are used in genetic engineering to 'cut' DNA in a specific place. Explain why the DNA needs to be cut. Use the diagram to help you.

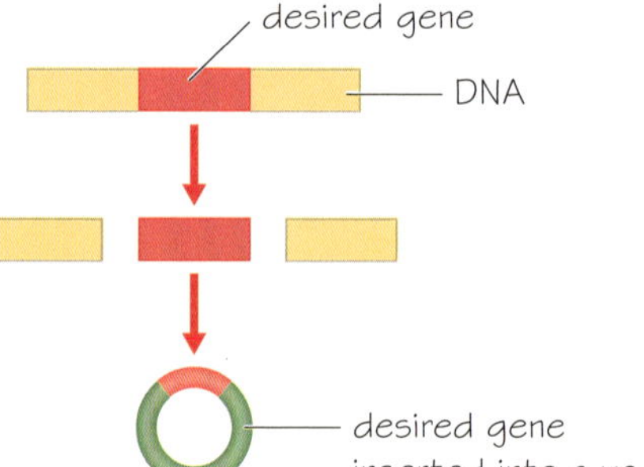

New genes are often inserted at an early stage of an organism's development.

Remember Genes are small sections of DNA that code for a characteristic.

Skills boost

1 How do I explain the impact of selective breeding?

Selective breeding changes the number of alleles in a population. There are benefits and risks linked to this change.

Variation in characteristics within a population is caused by the different versions of the genes that code for the physical characteristic. For example, the gene for stem length might have a long and a short stem version. The different versions of the same gene are called alleles.

1 Draw lines to link the type of population and the variation in alleles to the resulting impact on variation.

Variation is the differences in characteristics of organisms.

Type of population	Variation in alleles	Impact on variation
wild population	fewer alleles	more variation in characteristics
selectively bred population	many different alleles	less variation in characteristics

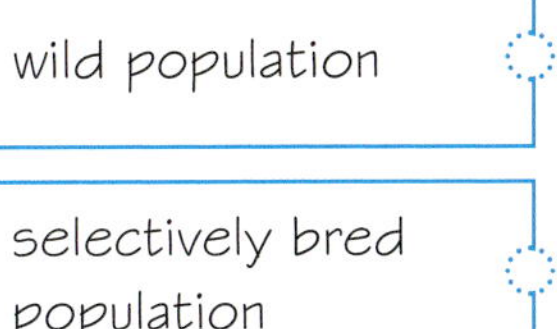

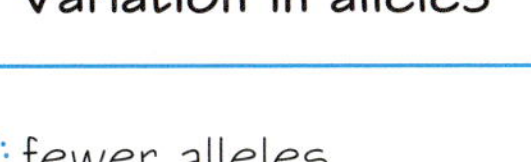

2 Which of the following statements best explains why there are normally fewer alleles in a selectively bred population?

Tick **one** box.

A Because they have fewer genes in their DNA ☐

B Because parents that live the longest are bred together ☐

C Because only the parents with the alleles for the desirable characteristics are bred together ☐

Resistance to disease may be inherited. Some organisms have alleles that provide resistance to a particular disease.

3 Explain why a population of selectively bred plants is less likely to have resistance to a disease than a wild population.

....

....

4 Circle the correct words in bold to summarise the effects of fewer alleles in a population.

Remember Fewer alleles means the population will be more closely genetically related so likely to be susceptible to similar diseases.

Fewer alleles in population →

- Population is **more / less** likely to be prone to the same disease
- **Increased / decreased** chance of inheriting genetic conditions or defects caused by homozygous recessive genes

Remember Homozygous is when both alleles for a gene are the same. Heterozygous is when both alleles for a gene are different.

Many genetic problems are caused by recessive alleles. These can stay hidden in a population because organisms can be carriers but not have the genetic defect. However, if two heterozygous carriers are bred together they have a much higher probability (25%) of having offspring with the defect.

Biology

How do I describe the stages of genetic engineering?

Genetic engineering is the process of modifying the genome of an organism. This is done by introducing a gene from another organism to give a desired characteristic. This process is used to produce useful substances such as insulin and to modify organisms such as plants so they have desirable characteristics.

The diagram shows the stages in genetically engineering bacteria to produce insulin that is used to treat people with diabetes.

Genome: all the genetic material in an organism.

Bacteria often have 'extra DNA' in their cells that look like a hoop of DNA. These are called **plasmids**. Scientists use them in genetic engineering because they are easy to cut open and change.

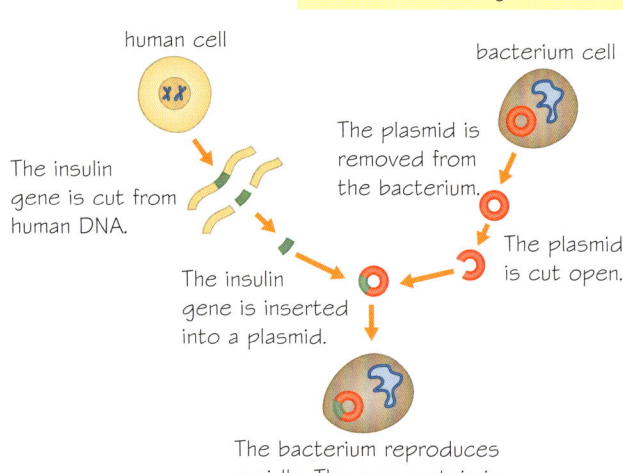

human cell

bacterium cell

The insulin gene is cut from human DNA.

The plasmid is removed from the bacterium.

The plasmid is cut open.

The insulin gene is inserted into a plasmid.

The bacterium reproduces rapidly. The new protein is produced by bacteria.

1. Add a label to the bacterium cell in the diagram to show the plasmid.

Restriction enzymes are used in genetic engineering to:

- cut out the gene for the desired characteristic in the DNA of an organism and expose its '**sticky ends**'
- cut open the bacterial or viral plasmid.

Ligase enzyme is used to join the two pieces of DNA back together. This is possible because they both have matching sticky ends.

Ligase gets its name from a 'ligature', which is a piece of cord that is used to tie something very tightly.

2. Circle the labels on the diagram to show where enzymes are used.

Genetic engineering requires a vector. A vector is something that is used to carry genetic material from one organism to another.

3. What is the vector used in the example shown in the diagram? Underline the correct answer in the box below.

   the human cell the bacterium cell the plasmid the enzymes

4. Complete the table by adding numbers to show the order of the main stages of genetic engineering.

Desired gene is inserted into plasmid using ligase enzyme. It fits perfectly as both pieces of DNA have matching sticky ends.	
Desired gene is cut out of organism with desired characteristic using restriction enzymes, leaving strands of DNA with 'sticky ends'.	
Plasmid containing desired gene is put into target organism using a vector.	
Plasmid is cut open using the same restriction enzymes to produce the same sticky ends.	

How do I explain the benefits and risks of selective breeding and genetic engineering?

To explain the potential benefits and risks of genetic engineering, you need to be familiar with some of the issues surrounding the use of genetically modified (GM) crops.

1. Draw lines to link the type of GM crop to its impact.

GM crop	Impact
A insect-resistant crop	a crop can be grown successfully in hot and dry areas
B drought-resistant crop	b less pesticides need to be applied so there is less pollution
C herbicide-resistant crop	c herbicide can be applied to kill weeds without damaging crop

2. Crops such as wheat can be genetically engineered to be resistant to herbicides. The diagram shows a possible effect of growing wheat that has been genetically engineered to be resistant to herbicide.
Answer the following questions.

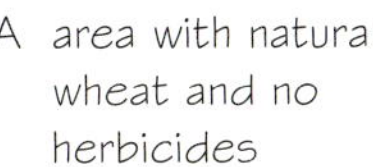

A area with natural wheat and no herbicides

B area with GM wheat where herbicides are used

Herbicides are substances that kill weeds. If a crop is resistant to herbicides, it means the farmer can apply herbicides to kill weeds without damaging the crops.

a What do weeds provide for organisms like small animals and insects? ....

b Predict the effect on these organisms when the weeds are killed. ....

c What would happen to the size of the insect and small animal populations if the weeds are killed? ....

d Fewer insects also means that fewer plants can be pollinated by insects. What would be the effect of this on the reproduction of wild plants? ....

e Which area, A or B, is likely to be able to support a wider range of species? ....

Biodiversity is the number of different species living in an area.

Biodiversity often decreases in areas with herbicide-resistant GM crops.

f Which area, A or B, do you think will have the most biodiversity? ....

Weeds compete with crops for water, light and minerals. By removing or reducing weeds, the farmer can increase his crop yield.

3. Complete the the sentences explaining the benefits of growing herbicide-resistant GM wheat plants.

Less herbicide needs to be used because the GM crop is ....

There is increased crop yield because the herbicide kills the weeds so there is less competition for resources, such as ....

Farmers make more profit because yield is increased and there are fewer costs, such as ....

Remember If you are asked to explain a risk or a benefit you must say why it happens. Linking words such as 'so', 'because' or 'therefore' are a good way of doing this.

Sample response

Your understanding of genetic engineering is likely to be tested in the context of bacteria that have been modified to produce useful substances, or crops or animals that are bred for desirable characteristics. Read the questions carefully, use your knowledge and consider your response.

Look at this exam-style question and student response.

Exam-style question

1 Bacterial cells can be genetically modified using enzymes to produce clotting proteins to treat people with haemophilia. Haemophilia is a genetic disorder that impairs the body's ability to make blood clots.

The human clotting gene is inserted into bacteria so that they contain the gene to produce the clotting protein.

(a) Explain how enzymes are used to produce a genetically modified plasmid in a bacterial cell. **(3 marks)**

(b) The bacteria used in this process is described as a 'vector'. Explain what is meant by a 'vector' as used in genetic engineering. **(2 marks)**

(a) *The blood clotting gene is isolated using restriction enzymes that cut the DNA.*
The circular bacterial plasmid is then cut open using the same restriction enzymes.
The gene can then be inserted into the plasmid of the bacteria using ligase enzyme.
The bacteria are grown in large tanks.
The bacteria then produce the blood clotting protein which can be extracted.

(b) *Something that carries a gene.*

1 The student's answer to question 1 (a) contains **two** sentences that would not gain any marks because they are giving information that is not relevant to the question.

a Highlight the **three** sentences that answer the question.

b Underline the **two** sentences that do not answer the question.

c The student's answer to question 1 (b) is missing some information. Write an improved definition of a vector.

....

....

Your turn!

It is now time to use what you have learned to answer the exam-style question from page 33. Remember to read the question thoroughly, looking for information that may help you. Make good use of your knowledge from other areas of biology such as ecology and genetics.

Exam-style question

1 Scientists have used selective breeding to produce a new variety of wheat plant that is better quality and produces a higher yield than wild wheat plants.

The new variety of wheat plant has fewer different alleles than wild wheat plants.

(a) Explain why the process of selective breeding causes a reduction in the number of different alleles in the population of wheat plants. **(2 marks)**

Remember If a farmer breeds from the best plants, they are likely to have similar characteristics and therefore similar alleles.

....

....

....

....

(b) Give **one** reason why it can be a disadvantage for a variety of wheat to have fewer different alleles. **(1 mark)**

Think about the effect of a smaller range of different alleles on the chances of an organism having an allele that gives resistance to a disease.

....

Wheat plants can be genetically engineered. This type of wheat is called genetically modified (GM) wheat. Some types of GM wheat are resistant to herbicides.

(c) Give **two** reasons why a farmer might benefit from growing GM wheat. **(2 marks)**

....

....

(d) Some people are concerned about the use of GM wheat. Suggest **two** reasons why. **(2 marks)**

....

....

....

....

Get back on track

Need more practice?

Exam questions may ask about different parts of a topic, or parts of more than one topic. Questions about selective breeding or genetic engineering could occur as:

- questions about selective breeding or genetic engineering only
- part of a question on the use of genetic modification in farming or biotechnology.

Have a go at these exam-style questions. 

Exam-style questions

1 Spider silk, used by spiders to make webs, is made from protein. It is very strong and flexible.

Scientists can now genetically engineer bacteria to produce the protein found in spider silk.

(a) State what is meant by the term 'genetic engineering'. **(1 mark)**

....

(b) Explain how scientists genetically engineer bacteria to produce the protein found in spider silk. **(4 marks)**

....

....

....

2 Crops can also be genetically engineered. Genetically modified (GM) crops have been genetically engineered to be resistant to some insects.

(a) Give **two** possible benefits of genetically engineering crops to be resistant to some insects. **(2 marks)**

....

(b) Give **two** possible risks of growing crops that have been genetically engineered to be resistant to some insects. **(2 marks)**

....

Boost your grade

To improve your grade, make sure you can:

- evaluate the risks and benefits of using selective breeding and genetic engineering to produce new varieties or breeds
- explain the difference in the stages involved in producing useful products from genetically modified bacteria compared with producing GM crops.

How confident do you feel about each of these **skills**? Colour in the bars.

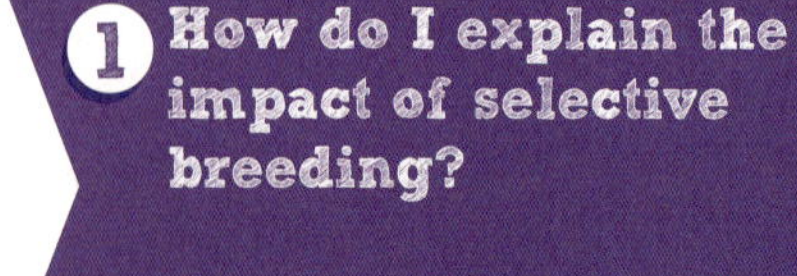

1 How do I explain the impact of selective breeding?

2 How do I describe the stages of genetic engineering?

3 How do I explain the benefits and risks of selective breeding and genetic engineering?

 Calculations in Biology

This unit will help you to show your understanding of the **quantitative units** used in biology, and how to convert between them. This will help you calculate the **actual size** of very small objects. It will also help you to represent very large or very small numbers using **standard form**.

In the exam you will be asked to answer questions such as the one below.

Exam-style question

1 A student looked at some onion cells using a light microscope.

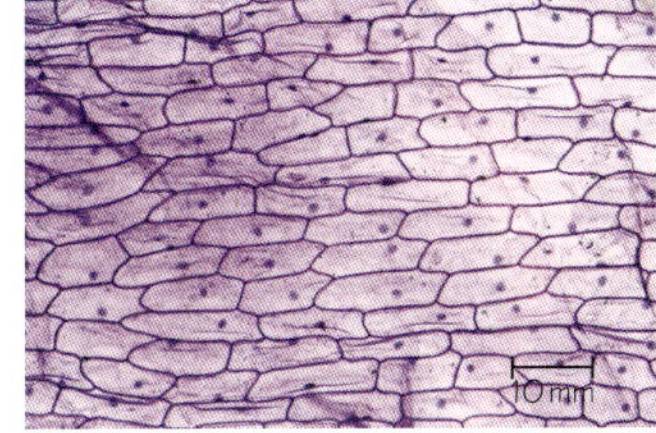

Magnification ×25

(a) The scale bar shows that the length of the cell in the image is 10 mm. The image has been magnified ×25 times. Calculate the **actual size** of the cell in μm. Show your working. **(2 marks)**

Actual size .... μm

(b) Mitochondria are organelles inside the cell which release energy for respiration. A typical mitochondrion is 0.003 mm. Write this size in **standard form**. **(2 marks)**

....

(c) On average there are 3×10^3 mitochondria in a typical onion cell.

Calculate how many mitochondria are in a sheet of onion skin containing 4×10^4 cells. Write your answer in standard form. **(2 marks)**

....

You will already have done some work on this topic. Before starting the **skills boosts**, rate your confidence in maths skills needed in biology. Colour in the bars.

1 How do I do calculations using numbers in standard form?

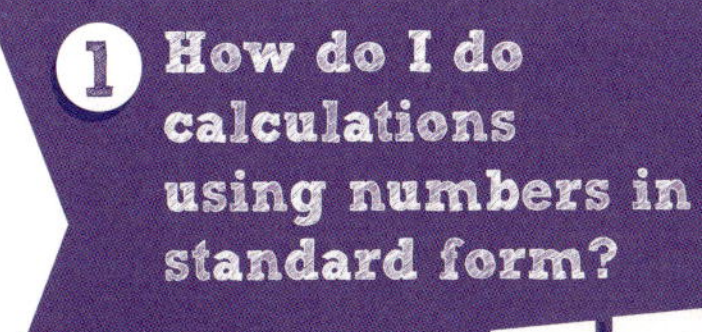

2 How do I convert between units?

3 How do I calculate the actual size of very small objects?

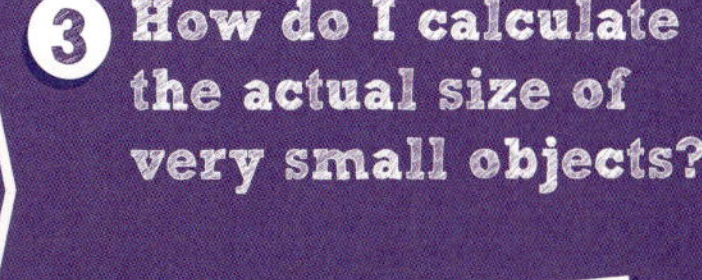

Get started

Standard form is a useful way of writing very large or very small numbers without writing lots of zeros. **7 000 000** is written as **7×10^6** and **0.000 008** is written as **8×10^{-6}**.

1 Complete the table, writing the ordinary numbers from their standard form.

Standard form	Ordinary number	Standard form	Ordinary number
2×10^5	200 000	6×10^0	
5×10^4		1×10^{-1}	0.1
3×10^3		8×10^{-2}	
4×10^2		9×10^{-3}	
7×10^1	70	4×10^{-4}	

Converting units is essential when measuring very small objects that can only be seen using a microscope. The diagram shows how the units relate to each other.

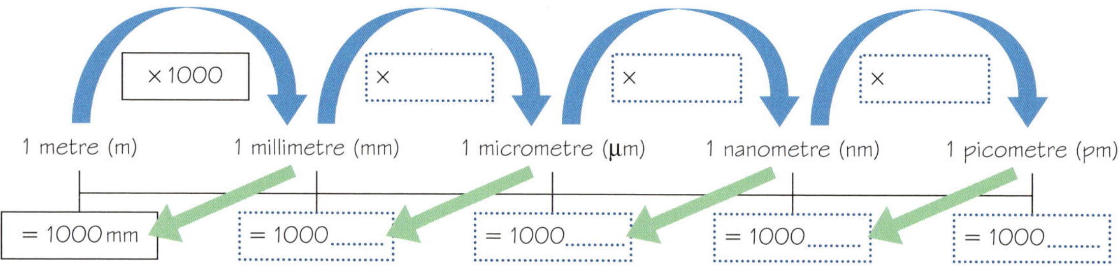

2 The conversion factor from m to mm is ×1000. Complete the boxes to show the other conversion factors and the correct units.

Learn what milli, micro, nano and pico mean.

3 Use the diagram above and the table below to complete the sentences that follow.

metre (m)	millimetre (mm)	micrometre (μm)	nanometre (nm)	picometre (pm)
	= 0.001 m	= 0.000 001 m	= 0.000000001 m	= 0.000 000 000 001 m
= 1000 mm		= 0.001 mm	= 0.000 001 mm	= 0.000 000 001 mm
= 1 000 000 μm	= 1000 μm		= 0.001 μm	= 0.000 001 μm
= 1 000 000 000 nm	= 1 000 000 nm	= 1000 nm		= 0.001 nm
= 1 000 000 000 000 pm	= 1 000 000 000 pm	= 100 000 pm	= 1000 pm	

a A metre (m) is 1000 times larger than a .... (mm)

b A micrometre (μm) is 1 000 000 times smaller than a .... (..............)

c A micrometre (μm) is 1000 times larger than a .... (..............)

d A picometre (pm) is 1 000 000 000 times smaller than a .... (..............)

To calculate the **actual size** of a very small object you have to rearrange the magnification equation.

$$\text{magnification} = \frac{\text{image size}}{\text{actual size}}$$

4 The image size of a cell is 50 mm. The actual size is 0.05 mm. Calculate the magnification.

a Write down the image size and actual size. image size .... actual size ....

b Substitute the values into the equation: magnification = $\frac{\text{..........}}{\text{..........}}$

c Write down the answer. magnification = × ....

Skills boost

How do I do calculations using numbers in standard form?

The method for writing numbers in standard form is shown in the diagram.

A is between 1 and 10 → $A \times 10^n$ ← n (index number) is a power of 10

If you want to write a large number in standard form, move the digits to the right until you have a number between 1 and 10. The number of moves is the power of 10. Take a look at the example.

$8400 = 8.4 \times 10^3$ (moves: 1 2 3)

1 An enzyme catalyses the reaction of 980 000 substrate molecules per minute. Write this in standard form.

a Move the digits to the right until you have a number between 1 and 10. ....

b How many moves did you make to the right? ....

c Write the new number $\times 10^{(\text{number of moves to the right})}$. ....

If you want to write a small number in standard form, move the digits to the left until you have a number between 1 and 10. The number of moves is the negative power of 10. Take a look at the example.

$0.0032 = 3.2 \times 10^{-3}$ (moves: 3 2 1)

2 A leaf palisade cell is 0.02 mm in length. Write this in standard form.

a Move the digits to the left until you have a number between 1 and 10. ....

b How many moves did you make to the left? ....

c Write the new number $\times 10^{-(\text{number of moves to the left})}$. ....

Sometimes you need to multiply numbers with powers. The diagram shows you how to do that.

$$(4 \times 10^3) \times (3 \times 10^4) = 4 \times 3 \times 10^3 \times 10^4 = 12 \times 10^7 = 1.2 \times 10^8$$

(multiply; add the powers; standard form)

3 A genetically modified bacterial cell produces 8×10^6 µg of insulin per hour. Calculate the total mass of insulin produced per hour by 3×10^5 bacterial cells.

Leave the numbers in standard form.

a Write out the calculation. ....

b Rewrite the multiplication, grouping the numbers and the powers of 10. Fill in the numbers.

$3 \times \square \times 10^{\square} \times 10^{\square} = \square \times 10^{(5+\square)} = \square \times 10^{\square} = \square \times 10^{\square}$ µg of insulin per hour

To multiply powers you add.

Write 24 in standard form.

The diagram shows you how to divide numbers with powers.

$$\frac{(2 \times 10^6)}{(4 \times 10^2)} = 2 \div 4 \times 10^6 \times 10^2 = 0.5 \times 10^4 = 5 \times 10^3$$

(divide; subtract powers; standard form)

4 A heart contains 3×10^{12} cardiac muscle cells. The total number of mitochondria in the heart is 1.2×10^{16}. Calculate the mean number of mitochondria in each cardiac muscle cell. ....

a Write out the calculation. $\frac{(\ldots\ldots\ldots\ldots\ldots\ldots)}{(\ldots\ldots\ldots\ldots\ldots\ldots)}$

There is no need to write them as ordinary numbers first.

b Rewrite the division. Group the numbers and the powers of 10. Fill in the numbers.

$\frac{1.2}{\square} \times \frac{10^{\square}}{10^{12}} = \square \times 10^{(\square - \square)} = \square \times 10^{\square} = \square \times 10^3$

To divide powers you subtract.

Put in standard form.

Biology

2 How do I convert between units?

When looking at cells, you need to demonstrate an understanding of the relationship between the quantitative units used to measure very small objects.

1 nm = 1000 pm, so multiply by 1000 to convert from nm to pm.

1 m = 1000 mm, and 1 mm = 1000 µm, so multiply by 1 000 000 to convert from m to µm.

1 Fill in the blank spaces on the diagram.

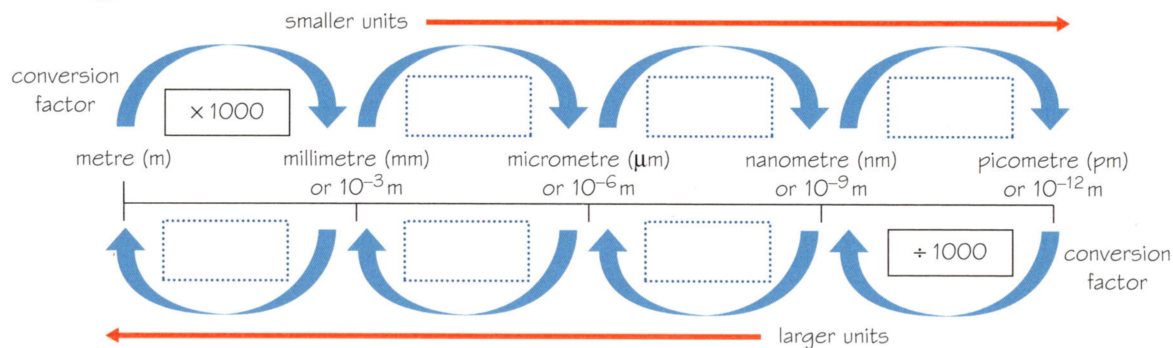

2 A student calculates a cheek cell under the microscope to be 60 µm long. Convert to nanometres (nm).

a What do you need to multiply the value by? ....

1000 nm = 1 µm.

b Do the calculation and write down the answer. .... nm

c Are nanometres (nm) larger or smaller than micrometres (µm)? ....

3 An image of a cell measures 35 mm across. Convert the measured image size from mm to nm.

a What do you need to multiply the value by? ....

1 000 000 nm in 1 mm.

b Do the calculation and write down the answer. .... nm

c Are millimetres larger or smaller than nanometres? ....

4 a Convert 620 000 000 pm to metres (m). Show your working out.

1 000 000 000 000 pm = 1 m.

b Tick the correct answer.

0.000 62 ☐ 0.62 ☐ 620 ☐ 620 000 ☐

5 A light microscope produces a resolution of 0.001 mm. Calculate this resolution in µm, nm and pm.

.... µm .... nm .... pm

6 Complete the table for the different organelles.

Organelle	Size (mm)	Size (µm)	Size (nm)
Chloroplast	0.006	6	
Mitochondrion	0.0004		
Ribosome	0.00022		
Nucleus			500 000

3 How do I calculate the actual size of very small objects?

Scientists often use µm or nm to compare the sizes of microscopic objects more easily.

Actual size is calculated by rearranging the magnification equation:

so $\text{magnification} = \frac{\text{image size}}{\text{actual size}}$ $\text{actual size} = \frac{\text{image size}}{\text{magnification}}$

The actual size will always be smaller than the image size.

1 The image size of a red blood cell is 8 mm. The magnification is ×40. Calculate the actual size.

a Substitute the values into the equation and calculate the answer.

actual size = (..............................) / (..............................) = ... mm.

b Write your answer to part **a** in standard form. ....

It will have a negative power.

c What is the actual size in micrometres (µm)? Tick the correct answer.

20 µm ☐ 250 µm ☐ 320 µm ☐ 200 µm ☐

2 The image size of a root hair cell is 9×10^{-3} m. The magnification is ×40.

a Calculate the actual size in mm.

First convert the image size to mm.

actual size = (..............................) / (..............................) = ... mm.

Multiply mm by 1000 to convert to µm.

b Now convert your answer to micrometres. Tick the correct answer.

225 µm ☐ 36 000 µm ☐ 360 µm ☐ 2250 µm ☐

3 An electron micrograph of a bacterial cell measures 3.6×10^{-2} m across. The magnification is ×6000. Calculate the actual size of the cell in metres (m).

....

Tick the correct answer.

Convert the standard form to an ordinary number first. Divide by the magnification and write the answer in standard form.

6×10^{-6} m ☐ 5×10^{-3} m ☐ 4×10^{-9} m ☐ 3×10^{-12} ☐

4 The width of the image of a nerve cell is 5 mm under an electron microscope with a magnification of ×20 000. Calculate the actual width of the nerve cell in mm. Write your answer in standard form.

It will be a number between 1 and 10 with digits moved to the left so a negative power.

Biology

Sample response

You may be asked to carry out several calculations based on practical work. Calculations can also be part of a theory question or use data provided on microorganisms.

Look at this exam-style question and student response.

Exam-style question

1 *Clostridium tetani* is a bacterium that causes an illness called tetanus. The photo shows some of these bacteria seen through a light microscope.

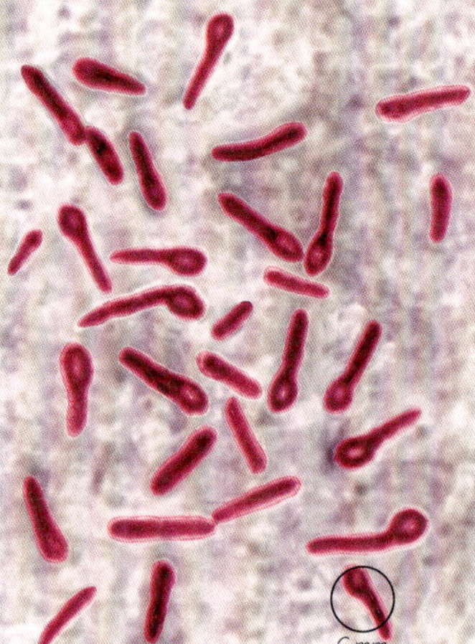

Magnification ×1000

(a) Calculate the actual length of this bacterial cell in μm. **(2 marks)**

$$\frac{1000}{6} = 166.7\,mm$$

(b) *Clostridium* bacteria reproduce rapidly. The population was estimated at 4 200 000 cells. Write this in standard form. **(2 marks)**

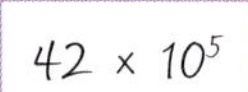

(c) A patient has 8.5×10^8 *Clostridium* bacteria spread across his skin. The surface area of his body is $1.7 \times 10^6\,mm^2$. Calculate the number of bacteria per mm^2 of skin. Write your answer in standard form.

$$(1.7 \times 8.5) \times 10^8 \times 10^6 = 14.45 \times 10^{14}$$

(2 marks)

1 **a** The student did not gain any marks in part **(a)**. What errors has the student made? 

....

b What should the correct answer be? 

....

2 For part **(b)** the student got 1 mark for the correct value but the answer is not in the correct standard form. How should the answer be written?

....

3 For the answer to part **(c)**, the student multiplied the two values together instead of dividing and didn't gain any marks. What should the correct calculation look like?

The answer should be smaller than the initial values.

Remember to subtract the powers.

Your turn!

It is now time to use what you have learned to answer the exam-style question from page 41. Remember to read the question thoroughly, looking for information that may help you. Make good use of your knowledge from other areas of biology.

Exam-style question

1 A student looked at some onion cells using a light microscope.

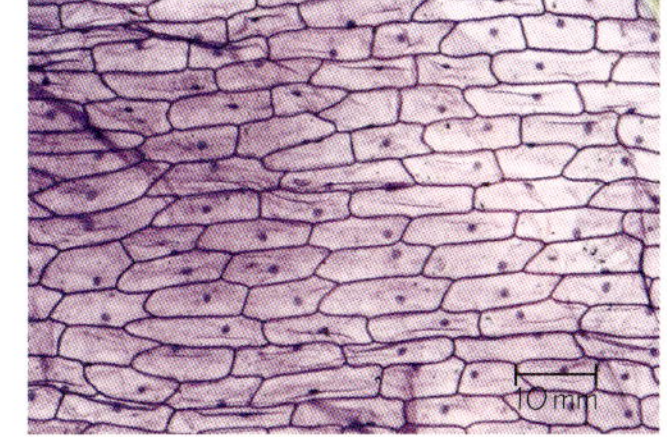

Magnification ×25

(a) The scale bar shows that the length of the cell in the image is 10 mm. The image has been magnified ×25 times. Calculate the **actual size** of the cell in μm. Show your working. **(2 marks)**

Actual size .... μm

Write out the equation first then substitute in the two values.

Look back at how to convert mm to μm: 1 mm = 1000 μm.

(b) Mitochondria are organelles inside the cell which release energy for respiration. A typical mitochondrion is 0.003 mm. Write this size in **standard form**. **(2 marks)**

Move the digits to the left until you have a number between 1 and 10. The number of moves is the negative power. 3 moves is 10^{-3}, 6 moves is 10^{-6} and so on.

(c) On average there are 3×10^3 mitochondria in a typical onion cell.

Calculate how many mitochondria are in a sheet of onion skin containing 4×10^4 cells. Write your answer in standard form. **(2 marks)**

Multiply the two standard form values together but add the powers.

Do you end up with another standard form or do you need to move the digits one more time and change the power?

Biology

Need more practice?

Exam questions may ask about different parts of a topic, or parts of more than one topic. Questions about use of correct units, standard form or actual size could occur as:

- part of a question on how to correctly apply these calculations
- part of a question about practical skills.

Have a go at these exam-style questions.

Exam-style questions

1 The diagram shows some specialised white blood cells called T helper cells. The image width of the labelled cell is 6 mm. The cells have been magnified 750 times. Calculate the actual width of this cell in µm. **(3 marks)**

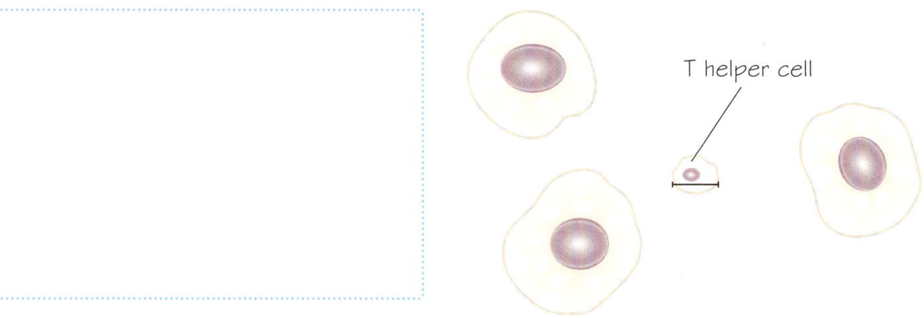

Read the question carefully and work out what you are being asked to do. Don't be put off by the context – you don't need to know about T helper cells to answer the question.

2 A sample of the patient's blood is 5×10^3 mm^3. It contains a count of 2.4×10^9 T helper cells.

Calculate the number of T helper cells per mm^3 of blood. Write the answer in standard form. **(3 marks)**

Boost your grade

Practise converting between units as many biology answers require this as a final stage in the calculation.

Standard form calculations will often be to do with numbers with lots of zeroes. Practise multiplying and dividing in standard form.

Microscope questions usually require magnification values but finding actual size is more challenging. Practise rearranging the magnification equation.

How confident do you feel about each of these **skills**? Colour in the bars.

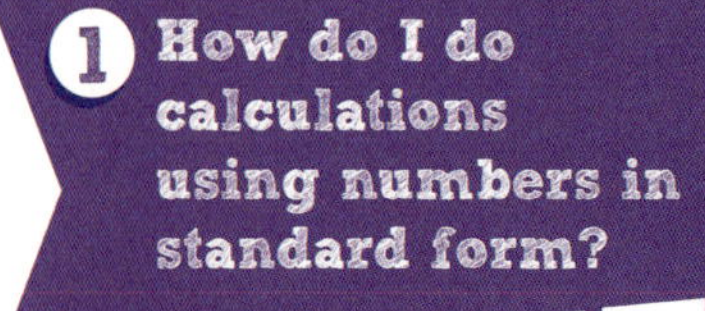

1 How do I do calculations using numbers in standard form?

2 How do I convert between units?

3 How do I calculate the actual size of very small objects?

Get started

AO2, AO3

Answering extended response questions

This unit will help you to understand what is needed for extended response questions by explaining command words so that you can work out what type of answer to give. This unit will also help you to plan and write your answers concisely.

In the exam, you will be asked to answer questions such as the one below.

Exam-style question

1 A new artificial heart is being trialled in patients with coronary heart disease.

Table 1 shows data that compares the performance of the new artificial heart design with the performance of the old artificial heart design.

		Patient numbers two years after the artificial hearts were implanted	
Design of heart	Total number of patients in study	Number of patients who are still alive	Number of patients who died
A: new design	90	63	27
B: old design	150	72	78

Table 1

Evaluate the data to decide which artificial heart design should be used in the future. **(6 marks)**

....

You will already have written some answers to extended response questions. Before starting the **skills boosts**, rate your confidence in your ability to understand, plan and write with the correct amount of detail the answer to an extended answer question. Colour in the bars.

1 How do I know what the question is asking me to do?

2 How do I plan my answer?

3 How do I choose the right detail to answer the question concisely?

Biology

Get started

The command word usually comes at the start of the question. It tells you what you need to do.

Here are some of the command words that are commonly found in extended response questions.

1 Draw a line to link each command word with its description.

The Edexcel website lists the command words you might need to use and what they mean.

Command word	Description
A Compare and contrast	a Say how or why something happens.
B Describe	b Use evidence from the information supplied to support an answer.
C Devise	c Use the information supplied as well as your own knowledge and understanding to consider evidence for and against and make a conclusion or judgement.
D Explain	d Describe the similarities and differences between things.
E Evaluate	e Plan a method or experiment using your knowledge.
F Justify	f Recall facts, events or processes in an accurate way.

Exam-style question

1 *In-vitro* fertilisation (IVF) is a process that can be used to help couples who are having difficulty conceiving a child.

Evaluate the use of IVF in treating infertility. **(6 marks)**

2 **a** Circle the command word in the exam-style question.

b What is the question asking you to do? Tick one box.

A Describe the process of IVF ☐

B Explain why couples use IVF ☐

C Give the risks and benefits of IVF ☐

When you are asked to evaluate something (or two things) you should weigh up the advantages and disadvantages or risks and benefits to come to a judgement.

Planning your answer is vital because the top marks are given for a sustained line of reasoning. This often means there should be a logical order to your answer.

3 Read these plan ideas and put them in a logical sequence by numbering the boxes. The first one has already been done for you.

After an introductory sentence, there should be the positives, then the negatives and then a general conclusion.

Even though there are risks involved, I think that if IVF will help a couple they should do it.	
IVF can be used to overcome blocked oviducts in a woman or a low sperm count in men.	
IVF does not always work first time and can take many cycles before a successful pregnancy.	
The embryo can be tested for genetic disorders.	
It can result in multiple births, which have a higher risk of miscarriage.	
IVF can overcome some reasons for infertility but it has risks and benefits.	1

1 How do I know what the question is asking me to do?

You can work out what the question is asking you by:

- looking at the command word, which tells you what to do
- identifying the topic and what aspect of the topic you are being asked about
- making sense of any information given in the question.

Some questions about genetic engineering are shown below.

Question	Answer content
A Describe the stages of genetic engineering.	a Give the reason why genetically engineered crops may have an impact on the number of species in an area.
B Explain the effect of genetically engineered crops on biodiversity.	b Weigh up the risks and benefits of genetically engineered crops.
C Compare and contrast genetic engineering with selective breeding.	c Give a brief account of each stage of genetic engineering.
D Evaluate the use of genetically engineered crops in farming.	d Give similarities and differences between genetic engineering and selective breeding.

1 a Underline the command word in each question.

b Draw lines to link each question to the correct answer content.

Exam-style question

1 The bacterium *Bacillus thuringiensis* produces a toxin that is extremely poisonous to certain species of insects. The gene that produces this toxin has been introduced into tomato plants. These genetically modified (GM) tomato plants have resistance to a range of insect pests that non-GM tomato plants do not have. This means that pesticides, which pollute the water and soil, do not need to be applied. However, some people are concerned that eating GM tomatoes might be bad for our health. Ecologists have also warned that the toxin might affect other useful insects that feed on tomato plants.

Evaluate the risks and benefits of growing tomatoes that contain the gene for the toxin. **(6 marks)**

2 a Circle the command word in the exam-style question above.

The command word often appears at the start of the question.

b What is the meaning of the command word?

....

c Which topic is the question testing? Tick one box.

genetic engineering ☐ selective breeding ☐

photosynthesis ☐ characteristics of bacteria ☐

d Highlight the words or phrases in the exam-style question that describe the scientific information you could use in your answer.

How do I plan my answer?

Take a short time to plan your answer. Decide on what to include from the topic and in which order to include it.

Exam-style question

1 Stem cells are used to treat some human diseases. **Table 1** gives information about using two types of stem cell to treat patients.

Stem cells from embryos	Stem cells from adult bone marrow
It costs £5000 to collect a few cells.	It costs £1000 to collect many cells.
There are ethical issues in using embryo stem cells.	Adults give permission for their own bone marrow to be collected.
The stem cells can develop into most other types of cell.	The stem cells can develop into only a few types of cell.
Each stem cell divides every 30 minutes.	Each stem cell divides every 4 hours.
There is a low chance of a patient's immune system rejecting the cells.	There is a high chance of a patient's immune system rejecting the cells.
More research is needed into the use of these stem cells.	Use of these stem cells is considered to be a safe procedure.

Table 1

Evaluate the use of embryonic and adult bone marrow stem cells for medical treatments. **(6 marks)**

....

To gain the highest marks, your answer needs to be in a logical order. This means that the information should be in an order that makes sense. For example, putting all the benefits then the risks of one form of treatment followed by the benefits then risks of the other treatment.

1 Organise the information you are given in Table 1 into groups.

a Highlight the benefits of using embryonic stem cells in one colour.

b Highlight the risks of using embryonic stem cells in another colour.

c Underline the benefits of using bone marrow stem cells.

d Circle the risks of using bone marrow stem cells.

2 Now plan your answer on a separate piece of paper, structuring your points in a logical order.

3 Complete the sentence below that weighs up the risks and benefits to make a judgement about which is better.

Check that all the points have been covered.

I consider that using .... stem cells is a better form of treatment because ....

A judgement is a personal opinion. Make sure you support your judgement with a reason. For example, 'I consider A is a better treatment than B **because** A's benefits outweigh the risks more than for treatment B'.

3 How do I choose the right detail to answer the question concisely?

You can get the right amount of detail in your answer by selecting the parts of the topic that answer the question rather than by attempting to write everything you know about the topic. Refer back to the command word to work out the style and content for your answer.

Exam-style question

1 Human embryonic stem cells are used for research into certain diseases and to find out how cells respond to potential new drugs. Many people are opposed to the use of embryonic stem cells on ethical grounds. Other people feel that there are good ethical reasons for the research.

Evaluate the use of embryonic stem cells in medical research. **(6 marks)**

1 Here are some student notes on embryonic stem cells. In the box beside each statement, write whether it is **for**, **against** or **irrelevant** to the argument.

Statement	
Scientists in many countries are using stem cells for research.	
Embryonic stem cells are potentially very useful because they have the ability to become any type of cell.	
Human embryos deserve respect, as does any human being.	
Unwanted embryos used in stem cell research are left over from fertility clinics. If not used they would be stored or destroyed.	
The two types of stem cells are embryonic stem cells and adult stem cells that are found in adult tissues.	

Some questions provide data in the form of a table or graph. Sometimes you will need to process the data to be able to effectively evaluate it.

When evaluating, only include information that highlights the positives and negatives that are important. General facts are irrelevant. Leave them out.

2 Evaluate the use of two different drugs trialled to treat coronary heart disease (CHD).

Drug	Total number of patients trialled	Number of patients needing surgery in 5 years	Total cost of drug trial
A	1200	66	£120 000
B	1000	58	£112 000

The total number of patients trialled is different, which makes comparison difficult. Some processing is needed to provide useful data.

a Which drug had the highest percentage of patients needing surgery after 5 years?

Drug A = $\left(\frac{66}{1200}\right) \times 100 =$ ....

Drug B = $\left(\frac{58}{1000}\right) \times 100 =$ ....

b Which drug works out the cheapest per patient?

Drug A = $\frac{£120\,000}{1200} =$ ....

Drug B = $\frac{£112\,000}{1000} =$ ....

c Based on your processed data, which drug would you say is most effective in your conclusion? Give a reason for your answer.

....

....

In this case, consider effectiveness and price, and refer to it in your judgement.

Get back on track

Sample response

Use this student response to improve your understanding. Consider what the graph is showing and use your knowledge to try to identify the issues involved. Think about what the question is asking and the science behind the data.

Look at this exam-style question and student response.

Exam-style question

1 **Figure 1** shows the relationship between body mass index (BMI) and the risk of developing type 2 diabetes for men and women.

Evaluate the link between BMI level and the risk of developing type 2 diabetes in men and women using evidence from the graph. **(6 marks)**

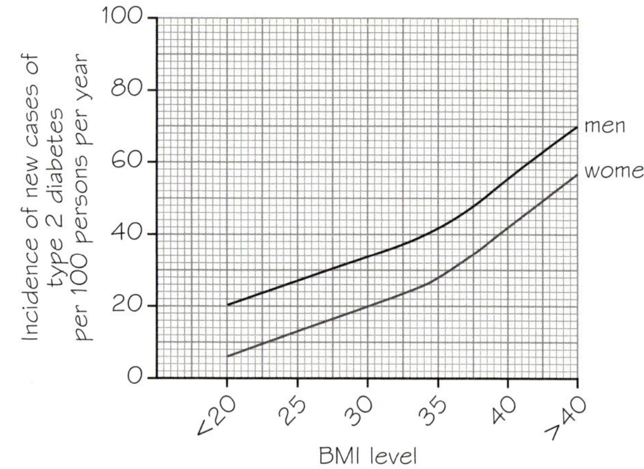

Figure 1 Obesity and diabetes risk

At greater than 40 BMI, there were 58 new cases per 1000 women per year. As the BMI goes up, the number of cases of type 2 diabetes also goes up. BMI means Body Mass Index and over 30 means that a person is obese. If they are obese, then they have a bigger chance of getting type 2 diabetes. At less than 20 BMI, there were 20 new cases per 1000 men per year. Men have a greater risk of developing type 2 diabetes than women because the line of their graph is higher than that of women.

1. Did the student successfully evaluate the link? **Yes / No**

2. Highlight where the student has made links between BMI and diabetes risk.

3. Cross out any information that is not relevant.

4. Circle where the student has used the data provided to support the answer.

5. What other additional or relevant information could the student have included that would improve the answer?

   ....

   ....

6. Does the response have a logical sequence? **Yes / No**

7. On a separate piece of paper write an improved answer.

In this question, the student has to compare the risk of men developing type 2 diabetes against the risk of women developing type 2 diabetes at different BMI values.

This is any information not used to compare men with women or form a conclusion.

Has the student processed useful data from the graph?

Has the student selected relevant detail?

Your turn!

It is now time to use what you have learned to answer the exam-style question from page 49. Remember to read the question thoroughly, looking for information that may help you. Make good use of your knowledge from other areas of biology.

Exam-style question

1 A new artificial heart is being trialled in patients with coronary heart disease.

Table 1 shows data that compares the performance of the new artificial heart design with the performance of the old artificial heart design.

		Patient numbers two years after the artificial hearts were implanted	
Design of heart	Total number of patients in study	Number of patients who are still alive	Number of patients who died
A: new design	90	63	27
B: old design	150	72	78

Table 1

Evaluate the data to decide which artificial heart design should be used in the future. **(6 marks)**

1 Circle the command word.

What does the command word mean? Keep that as the focus of your answer.

2 Processing this data to percentages will enable a useful comparison. Complete the table. The first one has been set up for you.

		Patient numbers two years after the artificial hearts were implanted	
Design of heart	Total number of patients in study	% of patients still alive	% of patients who died
A: new design	90	$\left(\frac{63}{90}\right) \times 100 =$	
B: old design	150	$\left(\frac{\ldots}{\ldots}\right) \times 100 =$	

3 Put these plan ideas in order by numbering the boxes. The first one has been done for you.

Survival rate of the old design after 2 years is 48%.	☐	*New design tested on a smaller sample.*	☐
Old design tested on a larger sample.	☐	*Comparison of survival rates.*	☐
Justified conclusion.	☐	*'Evaluate' so look at arguments for and against.*	1
Survival rate of the new design after 2 years is 70%.	☐	*Comparison of sample size.*	☐

4 Using the plan, write your own answer to the question. Use a separate sheet of paper.

Use the data to support your answer.

Biology

Need more practice?

Exam questions may ask about different parts of a topic, or parts of more than one topic. Questions about answering extended response questions could occur as:

- questions about that topic only
- part of a question on any topic you have studied, e.g. stem cells or genetic engineering
- part of a question about an experiment or investigation.

Have a go at this exam-style question. Write your answer on a separate sheet of paper.

Exam-style question

1 Fertility decreases with age as the ovaries stop releasing eggs.

In-vitro fertilisation (IVF) can be used to help women aged 50–60 years old to have a child.

A 55-year-old woman decided to try to have a child using IVF in 2017.

Table 1 shows the statistics for 2017 from the IVF clinic she attended.

Age range	30–39	40–49	50–59	60+
Number of IVF treatments	125	90	25	12
Average number of embryos transferred	2.1	2.7	3.1	3.7
Number of successful pregnancies	50	27	5	2

Table 1

Evaluate her decision. Use data from the table to support your answer. **(6 marks)**

The IVF process produces embryos '*in vitro*' or in a test tube. The embryos are then transferred back to the woman.

Remember to plan, answer to the command word and only include relevant information or processed data in a logical sequence.

You only have a very short time to plan, but the more you practise the faster you will become.

Boost your grade

Ensure you know the meanings of all the command words used for extended response questions.
Practise making your answers relevant by picking a question from sample assessment materials and writing the four most important points about the topic.

How confident do you feel about each of these **skills**? Colour in the bars.

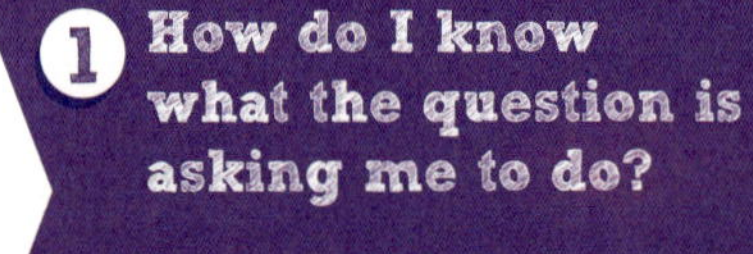

1 How do I know what the question is asking me to do?

2 How do I plan my answer?

3 How do I choose the right detail to answer the question concisely?

AO2

1 Moles

This unit will help you to describe what a mole is and how to do calculations involving moles. It will also help you calculate an empirical formula.

In the exam, you will be asked to answer questions such as the one below.

Exam-style question

1 Calcium carbonate, $CaCO_3$, decomposes to form calcium oxide and carbon dioxide when it is heated.

(a) Calculate the number of moles in 286 g of $CaCO_3$.

(relative atomic masses: Ca = 40, C = 12, O = 16)

number of moles = ... **(2 marks)**

(b) Calculate the number of molecules in 5 mol of carbon dioxide.

(Avogadro constant = 6.02×10^{23} mol^{-1})

number of molecules = ... **(1 mark)**

(c) A sample of calcium oxide contains 50 g of calcium and 20 g of oxygen by mass.

Calculate the empirical formula of calcium oxide.

empirical formula = ... **(3 marks)**

Chemistry

You will already have done some work on moles. Before starting the **skills boosts**, rate your confidence in each area. Colour in the bars.

1 How do I describe what a mole is?

2 How do I calculate how many moles, or particles, there are in a substance?

3 How do I calculate empirical formulae?

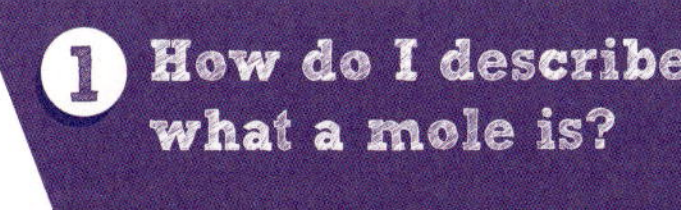

Get started

Moles are a way to consider the proportions, or ratios, of different substances involved in a reaction. A balanced equation shows the proportions of the substances in moles. The SI unit symbol for the mole is mol.

Consider that potential difference is measured in volts using the symbol V, e.g. a potential difference of three volts is 3 V. The amount of substance is measured in moles using the symbol mol, e.g. an amount of substance of three moles is 3 mol. We use 'mol' when a number is involved. For example, 1 mol of propane reacts with 5 mol of oxygen to produce 3 mol of carbon dioxide and 4 mol of water.

1 Complete the equation.

$$C_3H_8 + \text{..........} O_2 \rightarrow \text{..........} CO_2 + \text{..........} H_2O$$

Moles are also a way to consider the proportions of different particles within a substance.

For example:

1 mol of methane molecules, CH_4, contains 5 mol of atoms (1 mol of C atoms, 4 mol of H atoms).

1 mol of the ionic compound sodium chloride, NaCl, contains 1 mol of Na^+ ions and 1 mol of Cl^- ions.

2 Join each particle with the number of moles of it contained in calcium hydroxide, $Ca(OH)_2$.

Particle		Number of moles	
A	H atoms	a	1
B	Ca^{2+} ions	b	2
C	OH^- ions	c	3
D	O atoms	d	4

3 Write the number of moles of the different particles in 1 mol of ammonium hydroxide, NH_4OH.

In this compound, there are atoms and ions. Write down all the atoms and ions and how many there are of each.

....

....

4 Complete this sentence.

$Cu^{2+} + 2e^- \rightarrow Cu$

This reaction has .... mol of Cu^{2+} ions, ... mol of electrons and .... mol of Cu atoms.

You also need to use the following:

- The relative atomic mass (A_r) for an atom is given in the periodic table. It is the average mass of 1 mol of atoms of that element.
- The relative formula mass (M_r) of a compound, molecule or giant covalent substance is the sum of the relative atomic masses of all the atoms or ions in the formula.

Remember A giant covalent substance can be an element, such as diamond or graphite (which are forms of carbon). It can also be a compound, such as silica (which contains silicon and oxygen atoms).

Remember Do not confuse the quantity **mole** with the particle **molecule**.

How do I describe what a mole is?

You need to recall the definition of a mole and use it to describe quantities of substances.

One **mole** of particles of a substance is defined as either:

- the Avogadro constant number of particles (6.02×10^{23} atoms, ions, molecules or electrons)

Or:

- a mass of relative particle mass (g).

For example, each mole of methane, CH_4, contains 6.02×10^{23} CH_4 molecules.

In this definition, 'particles' covers atoms, ions, electrons, molecules and compounds.

1 Circle the A_r for lithium.

7 Li lithium 3

2 Fill in the blanks.

The relative particle mass could be:

- the relative .... mass, ..., found in the .... table
- the ...., M_r, calculated from .... values and the chemical formula.

If you are not given the A_r values, you can find them in the periodic table.

3 Calculate the relative particle mass of a sulfur dioxide molecule, SO_2.

A_r for sulfur, S, is 32

A_r for oxygen, O, is 16

The relative particle mass of SO_2 is .... + ... + ... = ...

So 1 mol of SO_2 molecules has a mass of .... g

4 Complete the table. Use a periodic table to help you.

Substance	Particle type	Formula	Relative particle mass	Mass of 1 mol (g)
Sodium chloride	ions	NaCl	23 + 35.5 = 58.5	58.5
Magnesium				
Water				
Glucose		$C_6H_{12}O_6$		
Sulfate	ion	SO_4^{2-}		
	atom		39	

Remember We can also have moles of electrons.

Chemistry

Skills boost

2 How do I calculate how many moles, or particles, there are in a substance?

You can calculate the **number of moles** in a substance if you know the mass of the substance. If you know the number of moles, you can calculate how many particles there are using the Avogadro constant.

You can use this equation to calculate the number of moles in a certain mass of a substance:

$$\text{number of moles} = \frac{\text{mass of substance (g)}}{A_r \text{ or } M_r}$$

Exam-style question

1 Calculate the number of moles in 185.25 g of $CuCO_3$.

(relative atomic masses: Cu = 63.5, C = 12, O = 16)

1 a Circle the value for mass in the question.

b The M_r of $CuCO_3$ is 123.5. Show how this is calculated.

....

....

c Substitute the values into the equation and calculate the answer.

$$\text{number of moles} = \frac{\text{mass (g)}}{M_r} = \frac{185.25}{\dots\dots} = \dots\dots \text{ mol}$$

Exam-style question

2 Calculate the number of moles of chlorine molecules in 213 g of chlorine.

(relative formula mass: Cl_2 = 71)

2 Highlight the values in the question. Calculate the answer.

$$\text{number of moles} = \frac{\dots\dots}{\dots\dots} = \dots\dots$$

To calculate the **number of particles** in a substance, multiply the number of moles by the Avogadro constant. (Avogadro constant = $6.02 \times 10^{23}\,\text{mol}^{-1}$.)

3 Use your answer to question 2 to calculate the number of molecules in 213 g of chlorine.

(relative formula mass = 17)

....

....

How do I calculate empirical formulae?

An empirical formula is the simplest whole number ratio of atoms or ions in a substance. It can be calculated from the masses of the elements it contains.

Remember to use a periodic table to find the A_r values you need.

Exam-style question

1 224 g of iron reacts with 96 g of oxygen. Calculate the empirical formula of the iron oxide made.

1 **a** Circle the mass of iron used.

b What is the A_r of oxygen? ....

c Complete the table.

	Iron (Fe)	Oxygen (O)
Mass (g)		96
Relative atomic mass, A_r	56	
$\frac{\text{mass}}{A_r}$ = number of moles	$\frac{\text{.........}}{56} = 4$	$\frac{96}{\text{.........}} = 6$
Find the simplest ratio of the number of moles. Divide by the smallest number	$\frac{4}{4} = 1$	$\frac{6}{4} = 1.5$
If needed, multiply by 2 to make the simplest ratio as whole numbers	2	3
Ratio of atoms	.... iron atoms for every .... oxygen atoms	
Empirical formula	Fe_2O_3	

We cannot have half an atom, so we need to change the highlighted numbers so they are both whole numbers.

You do not have to set your working out in a table, but it is good practice to lay it out in separate columns for each element.

Remember Write down the empirical formula when you have done your calculation.

Exam-style question

2 5.4 g of aluminium reacts with 21.3 g of chlorine. Calculate the empirical formula of the aluminium chloride made.

2 **a** Circle the masses in the question.

b Where will you find the values for A_r? ....

c Complete the calculation by writing out a table like the one in question 1.

Chemistry

Sample response

Use these sample responses to questions about moles to see some common errors made in these types of question. Look back at the skills boosts to help you.

Look at these exam-style questions and sample responses.

Exam-style question

1 Calculate the number of moles in 69 g of sodium carbonate, Na_2CO_3.
(relative atomic masses: Na = 23, C = 12, O = 16)

The student gave the following answer. An appropriate number of significant figures has been used.

$M_r = 23 + 12 + 16 = 51$

$\text{number of moles} = \frac{\text{mass}}{M_r}$

$= \frac{69}{51} = 1.3529 = 1.4\,\text{mol}$

1 **a** Highlight the parts that are correct.

b Circle the parts that are wrong.

c Correct the answer.

Exam-style question

2 A sample of lead bromide contains 41.4 g of lead and 32 g of bromine by mass. Calculate the empirical formula of lead bromide.
(relative atomic masses: Pb = 207, Br = 80)

2 **a** Circle the mistakes the student has made in the table on the left.

b Complete the table on the right to show how to calculate it correctly.

Lead (Pb)	Bromine (Br)
$\frac{32}{270} = 0.12$	$\frac{41.4}{80} = 0.52$
$\frac{0.12}{0.12} = 1$	$\frac{0.52}{0.12} = 4.3$
Empirical formula is $PbBr_4$	

Lead (Pb)	Bromine (Br)
Empirical formula is	

Your turn!

It is now time to use what you have learned to answer the exam-style question from page 57. Remember to read the question thoroughly, looking for information that may help you. Make good use of your knowledge from other areas of chemistry.

Exam-style question

1 Calcium carbonate, $CaCO_3$, decomposes to form calcium oxide and carbon dioxide when it is heated.

(a) Calculate the number of moles in 286 g of $CaCO_3$.

(relative atomic masses: Ca = 40, C = 12, O = 16)

Which equation do you need to use?

What do you need to calculate first?

number of moles = ... **(2 marks)**

(b) Calculate the number of molecules in 5 mol of carbon dioxide.

(Avogadro constant = $6.02 \times 10^{23}\,mol^{-1}$)

number of molecules = ... **(1 mark)**

(c) A sample of calcium oxide contains 50 g of calcium and 20 g of oxygen by mass.

Calculate the empirical formula of calcium oxide.

Remember that it helps to set out separate columns for each element.

Empirical formula = ... **(3 marks)**

Need more practice?

Exam questions may ask about different parts of a topic, or parts of more than one topic. Questions about moles and empirical formulae could occur as:

- questions about that topic only
- part of a question on most chemistry topics, usually as a calculation
- part of a question about an experiment or investigation.

Have a go at these exam-style questions.

Exam-style questions

1 Calculate the number of moles of lithium ions in 282 g of lithium sulfate, Li_2SO_4.

(relative atomic masses: Li = 7, S = 32, O = 16)

Number of moles of lithium ions = ... **(3 marks)**

2 Calculate the number of hydroxide ions in 0.5 mol of sodium hydroxide, NaOH.

(Avogadro constant = $6.02 \times 10^{23}\ mol^{-1}$)

Number of ions = ... **(1 mark)**

3 A compound contains 360 g of carbon, 60 g of hydrogen and 1065 g of chlorine by mass. Calculate the empirical formula of this compound.

(relative atomic masses: C = 12, H = 1, Cl = 35.5)

Empirical formula = ... **(3 marks)**

Boost your grade

Practise making up your own calculations. Include compounds with more challenging formulae.

How confident do you feel about each of these **skills**? Colour in the bars.

1 How do I describe what a mole is?

2 How do I calculate how many moles, or particles, there are in a substance?

3 How do I calculate empirical formulae?

2 Chemistry calculations

This unit will help you to set out different chemistry calculations clearly and give your answers to an appropriate number of significant figures.

In the exam you will be asked to answer questions such as the one below.

Exam-style question

1 Ammonium sulfate is a salt used as a fertiliser. Ammonia solution reacts with dilute sulfuric acid to produce ammonium sulfate.

$$2NH_3(aq) + H_2SO_4(aq) \rightarrow (NH_4)_2SO_4(aq)$$

(a) Calculate the relative formula mass of sulfuric acid.

(relative atomic masses: N = 14, H = 1, O = 16, S = 32)

relative formula mass = ... **(1 mark)**

(b) Calculate the mass of ammonium sulfate produced when 3.4 g of ammonia reacts with excess sulfuric acid. Give your answer to 2 significant figures.

mass = ... g **(3 marks)**

Chemistry

You will already have done some work on chemistry calculations. Before starting the **skills boosts**, rate your confidence in each area. Colour in the bars.

1 **How do I set out calculations in a logical step-by-step way?**

2 **How do I give answers to an appropriate number of significant figures?**

3 **How do I calculate the mass of a reactant or product?**

Get started

'Calculate' means obtain a numerical answer, showing relevant working and including any units. Sometimes you will use an equation to calculate a numerical answer. The best approach is to answer the question in a systematic and organised way, showing your method step by step. You can also obtain marks for showing your understanding, even if you get the wrong answer.

Here is an example of an equation: number of moles $= \frac{\text{mass of substance (g)}}{A_r \text{ or } M_r}$

What if you know the number of moles and want to find the mass?

To rearrange equations of the type $A = \frac{B}{C}$:

Make B the subject

Multiply both sides by C: $A \times C = B \times \frac{\cancel{C}}{\cancel{C}}$

C's cancel out on the right-hand side: $B = A \times C$

Make C the subject

Multiply both sides by C: $A \times C = B \times \frac{\cancel{C}}{\cancel{C}}$

Divide both sides by A: $\cancel{A} \times \frac{C}{\cancel{A}} = \frac{B}{A}$

A's cancel out on the left-hand side. $C = \frac{B}{A}$

1 Tick the correct rearrangements of the equation:

concentration (g / dm^{-3}) $= \frac{\text{mass of solute (g)}}{\text{volume of solution (dm}^3\text{)}}$

mass of solute (g) $= \frac{\text{concentration (g / dm}^3\text{)}}{\text{volume of solution (dm}^3\text{)}}$ ☐

mass of solute (g) = concentration (g / dm^3) × volume of solution (dm^3) ☐

volume of solution (dm^3) $= \frac{\text{mass of solute (g)}}{\text{concentration (g / dm}^3\text{)}}$ ☐

volume of solution (dm^3) $= \frac{\text{concentration (g / dm}^3\text{)}}{\text{mass of solute (g)}}$ ☐

To rearrange equations of the type $A = B \times C$:

Make B the subject

Divide both sides by C: $\frac{A}{C} = B \times \frac{\cancel{C}}{\cancel{C}}$

C's cancel out on the right-hand side: $B = \frac{A}{C}$

Make C the subject

Divide both sides by B: $\frac{A}{B} = \cancel{B} \times \frac{C}{\cancel{B}}$

B's cancel out on the right-hand side. $C = \frac{A}{B}$

2 The equation for mass is mass of substance (g) = number of moles × A_r or M_r. Using the method above, write the equation for A_r or M_r.

....

You can also substitute the values into an equation and then solve it.

3 Calculate the mass of 0.6 mol of sodium using the equation: number of moles $= \frac{\text{mass of substance (g)}}{A_r}$ (A_r = 23)

a Substitute in the values. number of moles = ... $= \frac{\text{mass}}{\text{..............}}$

b Multiply each side of the equation by A_r. .... × .... $= \frac{\text{mass}}{\text{..............}}$ × ....

c Cancel the two values for A_r on the right-hand side of the equation and calculate the answer.

mass = ...

Use the method you prefer to make sure you can calculate what you need to.

Skills boost

How do I set out calculations in a logical step-by-step way?

It is important to set your calculations out clearly and show what you are doing step by step. You are more likely to get the correct answer, and it is also easier for you to check your method.

1. Here is a step-by-step checklist that will help you when doing a calculation. The steps are in the wrong order. Write the correct order in which the steps should happen.
   1. Calculate your answer.
   2. Substitute the correct values into the equation.
   3. Highlight important information given in the question, including any values.
   4. Calculate any values you need to put into the equation. Show your working out and convert any units if necessary (e.g. kg to g).
   5. Write down your answer to an appropriate number of significant figures and include units.
   6. Write down any equations you need to use and rearrange them if necessary.

   ....

Now try using the steps to answer the following questions.

2. Calculate the mass of 0.03 mol of methane molecules, CH_4.
   (relative atomic masses: C = 12, H = 1)
   1. Highlight the value for the number of moles you have been given in the question.
   2. Write down the equation and rearrange it: mass = number of moles × M_r
   3. You do not need to convert any units. Calculate the M_r value for CH_4.

      .... + (.................... × ....) = ...
   4. Substitute the values into the equation.

      mass = number of moles × M_r = ... × ....
   5. Calculate the mass. mass = ... g
   6. Check that your answer has the correct number of significant figures (in this case, 2 sf) and that the unit is given.

3. Calculate the number of moles in 0.31 kg of sodium oxide, Na_2O.
   (relative atomic masses: Na = 23, O = 16)
   1. Highlight the value for mass you have been given in the question.
   2. Write down the equation: number of moles = $\frac{\text{mass (g)}}{M_r}$
   3. Calculate the M_r value for Na_2O.

      (.................... × ....) + ... = ...

      Convert kg to g. 0.31 kg = ... g

      > To convert kg to g multiply by 1000.
   4. Substitute the values into the equation. number of moles = $\frac{\text{....................}}{\text{....................}}$ (g)
   5. Calculate the number of moles. number of moles = ...
   6. Check that your answer has the correct number of significant figures (in this case, 2 sf) and that the unit is given.

Chemistry

Skills boost

2 How do I give answers to an appropriate number of significant figures?

Using significant figures (sf) is a way of rounding to a given, or chosen, number of figures.

Sometimes you will be told how many significant figures to give your answer to. At other times you need to decide. You need to be able to round to 1, 2 or 3 significant figures.

The first significant figure is the first non-zero digit (1–9) on the left of any number. Zero can be a significant figure if it appears to the right of the first significant figure.

1 a Highlight the **first** significant figure in these numbers: 0.0801, 24.9

b Highlight the **second** significant figure in these numbers: 0.0801, 24.9

c Complete this statement.

37.4 has 3 significant figures: ...., ... and ....

When you round a number, if the next digit after the one you want is **5 or more**, round **up**. If the next digit is less than 5, do not round up.

d Complete these sentences by writing the correct answers.

If you write 37.4 to 1 significant figure, then the first sf is ....

The next digit is 7, so you round 3 up to .... You need to write a zero to keep the place value. So, 37.4 to 1 sf is ....

If you write 37.4 to 2 sf, then the second sf is .... The next digit is 4, so .... round up. 37.4 to 2 sf is ....

You may be told how many significant figures to give. If not, look at the number of significant figures used in any values you are given. Give your answer to the smallest number of significant figures in these values.

2 Write the number of significant figures each value has.

a relative atomic mass of copper, 63.5 ....

b mass of sodium chloride, 21 g ....

c a reaction time, 48.01 s ....

d moles of hydrochloric acid, 0.0033 mol ....

e moles of copper oxide, 5 mol ....

f Avogadro's constant, 6.02×10^{23} mol^{-1} ....

4 Complete this table. The first row has been done for you.

Number	To 1 sf	To 2 sf	To 3 sf
0.02564	0.03	0.026	0.0256
0.00083921			
1.035			
609.72			

3 How do I calculate the mass of a reactant or product?

If you know the mass of one substance in a reaction, you can calculate the number of moles. If you know the number of moles of one substance and the mole ratio, you can work out the number of moles of another substance. You can then calculate the mass of the other substance.

The mole ratio is the ratio of the number of moles of different substances shown in a balanced equation.

1 The displacement reaction between copper sulfate and excess magnesium produces magnesium sulfate.

$$Mg(s) + CuSO_4(aq) \rightarrow MgSO_4(aq) + Cu(s)$$

Calculate the mass of magnesium sulfate produced from 319 g of copper sulfate. Give your answer to 2 significant figures.

(relative atomic masses: Mg = 24, Cu = 63.5, S = 32, O = 16)

a Look at the information you are given in the question. Highlight the substances in the equation which are needed to answer the question.

b Write the relative formula mass of $CuSO_4$. ....

You need to calculate the number of moles of $CuSO_4$ so, using the balanced equation to find the mole ratio, work out the number of moles of $MgSO_4$. You can then calculate the mass of $MgSO_4$.

2 a Calculate the number of moles of $CuSO_4$.

number of moles of $CuSO_4$ = $\frac{\text{..............}}{\text{..............}}$

= ....

> To calculate the number of moles, use the equation:
>
> number of moles = $\frac{\text{mass of substance (g)}}{M_r}$

b Now consider $MgSO_4$. Write the relative formula mass of $MgSO_4$.

....

c Find the mole ratio.

From the balanced equation, 1 mol of $CuSO_4$ produces .... mol of $MgSO_4$.

This gives a mole ratio of 1 : ....

d How many moles of $MgSO_4$ are produced in the reaction? 

....

e Now calculate the mass of $MgSO_4$ produced.

mass = ...

> In this case, the number of moles of $MgSO_4$ produced in the reaction is the same as the number of moles of $CuSO_4$ that reacted.

> To calculate the mass, use the equation: mass of substance (g) = number of moles × M_r

Another way is to set out your answer in a table.

3 Complete the table.

Substance	$CuSO_4$	$MgSO_4$
Mass (g)	319	
M_r		
Mole ratio	1	
Number of moles		

> Check how many significant figures your answer is given to.

> Look at any balancing numbers in front of the substances.

Chemistry

Sample response

Use this student response to improve the way you answer calculations. Consider whether all the instructions have been followed and the working out has been shown in a step-by-step way.

Exam-style question

1 Calculate the mass of 1.68 mol of lead nitrate $Pb(NO_3)_2$. Give your answer to 3 significant figures.

(relative atomic masses: Pb = 207, N = 14, O = 16) **(2 marks)**

mass = moles × M_r

= 556.08 g

1 As you can see from the table, the student did not follow all the steps in the calculation.

	Has the student?	Have you?
Highlighted the important detail in the question.	no	
Shown the equation being used.	yes	
Shown the calculation for M_r.	no	
Substituted the values into the equation.	no	
Given the answer to the correct number of significant figures.	no	

Using the table as a guide, write a more detailed answer to the question. Tick the steps in the table as you do them.

Your turn!

It is now time to use what you have learned to answer the question from page 65. Remember to read the question thoroughly, looking for information that may help you. Make good use of your knowledge from other areas of chemistry.

Exam-style question

1 Ammonium sulfate is a salt used as a fertiliser. Ammonia solution reacts with dilute sulfuric acid to produce ammonium sulfate.

$$2NH_3(aq) + H_2SO_4(aq) \rightarrow (NH_4)_2SO_4(aq)$$

(a) Calculate the relative formula mass of sulfuric acid.

(relative atomic masses: N = 14, H = 1, O = 16, S = 32) **(1 mark)**

relative formula mass = ...

Highlight the information you need if that helps you.

Set out your working clearly to show how you get your answer.

(b) Calculate the mass of ammonium sulfate produced when 3.4 g of ammonia reacts with excess sulfuric acid. Give your answer to 2 significant figures. **(3 marks)**

mass = ... g

- What information have you been given?
- What do you need to calculate?
- What equation can you use?
- Set out your working clearly to show how you get your answer. Include all the steps.
- Have you given your answer to the correct number of significant figures?

Need more practice?

Exam questions may ask about different parts of a topic, or parts of more than one topic. Questions about chemistry calculations could occur as:

- questions about that topic only
- part of a question on most chemistry topics
- part of a question about an experiment or investigation.

Have a go at this exam-style question.

Exam-style question

1 Calculate the maximum mass of calcium nitrate that can be formed from 115 g of calcium carbonate and an excess of nitric acid. Give your answer to 3 significant figures.

(relative atomic masses: Ca = 40, C = 12, O = 16, H = 1, N = 14) **(2 marks)**

$$CaCO_3(s) + 2HNO_3(aq) \rightarrow Ca(NO_3)_2(aq) + H_2O(l) + CO_2(g)$$

mass = ... g

Boost your grade

Every time you do a calculation in chemistry, write down your working-out in a logical way. Use the checklist on page 71 to help you. Practise making up your own calculations to help your understanding.

How confident do you feel about each of these **skills**? Colour in the bars.

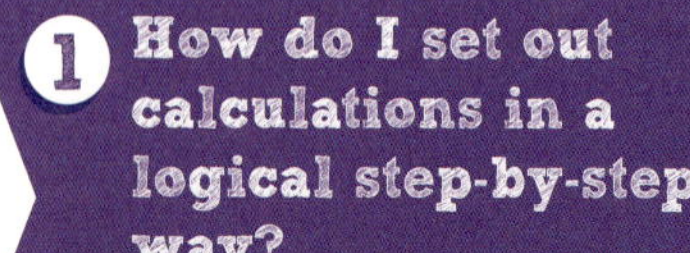

1 How do I set out calculations in a logical step-by-step way?

2 How do I give answers to an appropriate number of significant figures?

3 How do I calculate the mass of a reactant or product?

Get started

AO1, AO2

3 Chemical equations

This unit will help you to balance different chemical equations, and write ionic equations for reactions.

In the exam you will be asked to answer questions such as the ones below.

Exam-style questions

1 Carbon dioxide can be formed by the reaction of sodium carbonate, Na_2CO_3, with dilute nitric acid, HNO_3. Write the balanced equation for this reaction. **(3 marks)**

....

2 120 g of magnesium oxide, MgO, reacts completely with 378 g of dilute nitric acid, HNO_3, to form magnesium nitrate and water. Deduce the balanced equation for the reaction. You must show your working. **(4 marks)**

(relative atomic masses: Mg = 24, O = 16, H = 1, N = 14)

....

You will already have done some work on chemical equations. Before starting the **skills boosts**, rate your confidence in each area. Colour in the bars.

1 How do I balance chemical equations?

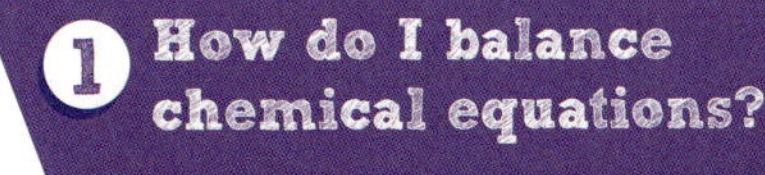

2 How do I balance an equation given the masses of reactants and products?

3 How do I write ionic equations given information about a reaction?

Get started

We write formulae to represent substances, especially in equations, and it is important to do it accurately.

> **Remember** Symbols must be given with the correct capital and lower case letters, and with capitals bigger than lower case. Any subscript numbers must be smaller than the letters and positioned just below the line the formula is written on.

1 Circle Ⓐ the correct formula for carbon dioxide.

CO2	CO^2	CO_2	CO_2	Co_2	co2

As there is always the same number of each atom before and after a reaction, add up the number of each atom on each side of the equation. You should count the atoms when balancing equations.

2 Complete the tables to show the number of each atom in each substance.

a $Al(OH)_3$

> The number 3 means there are 3 of each of the atoms inside the brackets.

Atom	Number
Al	
O	3
H	

> Brackets are only put around a group of atoms or an ion if there is more than one of them in the substance, e.g. NaOH does not need brackets around the OH ion.

b $2Li_2SO_4$

> The number 2 means there are 2 units of this substance.

Atom	Number
Li	4
S	
O	

State symbols are written in brackets next to the formula of a substance to show whether it is a solid, liquid, gas or solution. Give state symbols in equations when you are asked to.

3 Write the state symbols used in chemical equations.

> Aqueous (aq) means that the substance is dissolved in water.

solid .... liquid ....

gas .... aqueous ....

When you are given the mass of a substance, you can calculate the number of moles and use this to balance a chemical equation.

4 Complete the equation used to calculate the number of moles:

$$\text{number of moles} = \frac{\text{mass of substance}}{\text{..............................}}$$

5 Ionic equations show the ions involved in a reaction. Ions that do not change during a reaction are called spectator ions.

Balanced equation showing the reaction between magnesium and copper sulfate	Ionic equations showing what happens in the reaction
$Mg(s) + CuSO_4(aq) \rightarrow Cu(s) + MgSO_4(aq)$	$Mg(s) + Cu^{2+}(aq) \rightarrow Mg^{2+}(aq) + Cu(s)$

Write the spectator ion in the reaction. ....

How do I balance chemical equations?

A balanced chemical equation shows the chemical formulae of the reactants and products involved in a reaction. It has the same number of atoms of each element on either side of the arrow.

Follow these steps to balance a chemical equation:

1 Count how many atoms of each element there are in the reactants.
2 Count how many atoms of each element there are in the products.
3 If they are the same, the equation is balanced.
4 If they are not, then add a number at the front of the formula if more atoms are needed.
5 Re-count and repeat until the equation is balanced.

1 Here is the equation for the complete reaction between aluminium oxide and sulfuric acid:

	reactants		products
word equation	aluminium oxide + sulfuric acid	→	aluminium sulfate + water
formula equation	$Al_2O_3 + H_2SO_4$	→	$Al_2(SO_4)_3 + H_2O$

a Complete columns 1 and 2 in the table.

Element or group	Number of each atom, or group of atoms			
	1 Reactants	2 Products	3 Reactants	4 Products
Al	2			
O				
H	2			
S	1			
SO_4				

b Circle the correct words in bold to complete the sentences.

More oxygen atoms are needed on the **reactants / products** side.

More / less sulfur is needed on the reactants side.

More sulfate is needed on the **reactants / products** side.

Looking at groups of atoms, e.g. compound ions such as SO_4^{2-}, OH^-, NH_4^+, CO_3^{2-}, can be a good place to start.

c Start by writing a 3 in front of the reactant that contains SO_4^{2-} to balance this group.

Al_2O_3 + ... $H_2SO_4 \rightarrow Al_2(SO_4)_3 + H_2O$

d Complete columns 3 and 4 in the table above now.

e Which two atoms still need to be balanced? ....

f Write how many of each atom you need. ....

g Which product contains both of these atoms? ....

Check all of the atoms in every step to see if the number of each one has changed.

h Write the number you put in front of this product to get the extra atoms you need. ....

i Complete and check the balanced equation.

Al_2O_3 + ... $H_2SO_4 \rightarrow Al_2(SO_4)_3$ + ... H_2O

Remember A balanced equation shows the ratios of the substances in moles.

Chemistry

2 How do I balance an equation given the masses of reactants and products?

If you know the number of moles of some of the substances in a reaction you can work out the balanced equation from the mole ratio unless no substance is in excess. The number of moles can be calculated from the mass and A_r when it's an element (e.g. iron (Fe)), or M_r when it's a molecule (e.g. oxygen (O_2)) or a compound (e.g. carbon dioxide (CO_2)) of each substance.

Exam-style question

1 56 g of lithium reacts completely with 64 g of oxygen, O_2, to produce lithium oxide, Li_2O. Deduce the balanced equation for the reaction. **(3 marks)**

(relative atomic masses: Li = 7, O = 16)

Deduce means use information and show how you find it out.

In this example, you have been given the masses of the reactants to work out the balanced equation. You could also be given the masses of the products.

1 a Complete the table.

		Reactants	
		Li	O_2
Mass (g)	Find this information in the question		
A_r or M_r	Oxygen is O_2 – this means you need the M_r		
Calculate the number of moles	Number of moles = $\frac{\text{mass (g)}}{A_r \text{ or } M_r}$		
Mole ratio of the reactants	Write the simplest whole number ratio		

b Circle the correct numbers in bold in this sentence.

2 / **4** / **8** mol of Li react with **1** / **2** / **4** mol of O_2 to produce Li_2O.

It is not necessary to write 1 in front of a substance in the equation so we leave it blank.

2 Now balance the equation.

a Write the number of atoms in the reactants ($4Li + O_2$):

Li .... O ....

b Write the number of atoms in the product (Li_2O).

Li .... O ....

There are twice as many of each atom in the reactants as in the products so the equation can be balanced by putting a 2 in front of the product.

c Complete the balanced equation for the reaction.

.... Li + O_2 → .... Li_2O

Remember Make sure it is correct by checking there is the same number of each atom on each side.

How do I write ionic equations given information about a reaction?

Ionic equations show the ions that change during a reaction. Ions that appear on both sides of the reaction do not change so are not included in ionic equations. These are called spectator ions. Some ionic equations, called half equations, include electrons written as e^-.

1 Consider this neutralisation reaction.

word equation:	sodium hydroxide	+	hydrochloric acid	→	sodium chloride	+	water
balanced equation:	NaOH(aq)	+	HCl(aq)	→	NaCl(aq)	+	H_2O(l)

The ions in this reaction are shown in the box.

Na^+	OH^-	H^+	Cl^-	→	Na^+	Cl^-	H_2O

a Cross out the ions that appear on both sides of the equation. These are the spectator ions.

b Complete the ionic equation for this reaction. Include state symbols and ignore the spectator ions.

.... + ... → H_2O(l)

2 Consider this precipitation reaction.

potassium chloride	+	silver nitrate	→	potassium nitrate	+	silver chloride
KCl(aq)	+	$AgNO_3$(aq)	→	KNO_3(aq)	+	AgCl(s)

a List the four ions in the reaction.

....

b Cross out the spectator ions.

c Write the ionic equation for the formation of silver chloride. Include state symbols and ignore the spectator ions.

.... + ... → ....

A change of state from aqueous to solid shows which ions have taken part in the reaction.

3 Consider the electrolysis of a solution of copper(II) chloride.

Half equation at the negative electrode:

Cu^{2+}(aq) + $2e^-$ → Cu(s)

This shows that 1 mol of copper ions reacts with (gains) 2 mol of electrons.

Remember Each electron has one negative charge.

Complete the half equation at the positive electrode by balancing it:

.... Cl^-(aq) → Cl_2(g) + ... e^-

Cl_2(g) has no overall charge.

Sample response

Use this student response to improve the way you balance equations. Consider whether any working out has been shown in a step-by-step way.

Exam-style question

1 Sulfuric acid, H_2SO_4, reacts with lithium hydroxide, LiOH, to form lithium sulfate, Li_2SO_4, and water.

Write the balanced equation for this reaction. **(2 marks)**

$H_2SO_4 + Li_2OH \rightarrow Li_2SO_4$

1 a Highlight the substances in the question.

b Has the student written the correct formulae? **yes / no**

c Has the student included all of the substances? **yes / no**

d Write the unbalanced equation using the correct formulae.

....

e Complete the table.

Element	Number of each atom	
	Reactants	Products
H		
S		
O		
Li		

f Circle the elements in the table that are balanced.

g Highlight the elements that are not balanced.

h Is it a reactant, product or both that needs the number of atoms increasing?

....

i Write the balanced equation.

....

j Complete the table to check the equation is now balanced.

Element	Number of each atom	
	Reactants	Products
H		
S		
O		
Li		

Your turn!

It is now time to use what you have learned to answer the exam-style questions from page 73. Remember to read the question thoroughly, looking for information that may help you. Make good use of your knowledge from other areas of chemistry.

Exam-style questions

1 Carbon dioxide can be formed by the reaction of sodium carbonate, Na_2CO_3, with dilute nitric acid, HNO_3. Write the balanced equation for this reaction. **(3 marks)**

....

An acid reacts with a carbonate to form a salt, water and carbon dioxide.

When you have the correct formulae for reactants and products, check the equation is balanced.

2 120 g of magnesium oxide, MgO, reacts completely with 378 g of dilute nitric acid, HNO_3, to form magnesium nitrate and water. Deduce the balanced equation for the reaction. You must show your working. **(4 marks)**

(relative atomic masses: Mg = 24, O = 16, H = 1, N = 14)

....

....

....

....

- Show your working out step by step.
- Find the mole ratio of the reactants.
- Work out the formula for magnesium nitrate by counting the atoms on the reactants side of the equation, and think about the ions.
- Remember to write both products.
- Write out the full balanced equation.

Need more practice?

Exam questions may ask about different parts of a topic, or parts of more than one topic. Questions about chemical equations could occur as:

- questions about that topic only
- part of a question on any topic
- part of a question about an experiment or investigation.

Have a go at these exam-style questions.

Exam-style questions

1 When nonane undergoes complete combustion, a mixture of carbon dioxide and water is formed.

Complete the balanced equation for this reaction.

C_9H_{20} + ... O_2 → .... CO_2 + ... H_2O **(2 marks)**

2 Magnesium chloride solution, $MgCl_2$, reacts with sodium hydroxide solution, NaOH, to form a precipitate of magnesium hydroxide. Write the ionic equation for the formation of magnesium hydroxide. Include the state symbols. **(3 marks)**

....

Which ions form the precipitate?

Which are the spectator ions that do not go in the equation?

Check that the equation is balanced.

Boost your grade

Practise writing word equations as chemical equations and make sure they are balanced. Learn the formulae and charges for the ions you need to use.

How confident do you feel about each of these **skills**? Colour in the bars.

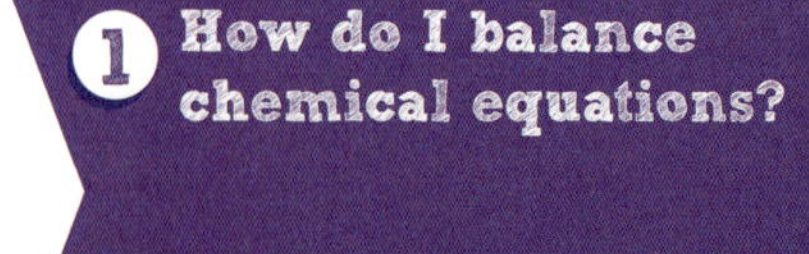

1 How do I balance chemical equations?

2 How do I balance an equation given the masses of reactants and products?

3 How do I write ionic equations given information about a reaction?

4 Dynamic equilibrium

This unit will help you to understand dynamic equilibrium. This unit will also help you to apply knowledge and understanding of scientific ideas, scientific enquiry, techniques and procedures.

In the exam, you will be asked to answer questions such as the one below.

Exam-style question

1 Dinitrogen tetroxide, N_2O_4, is a colourless gas.

Nitrogen dioxide, NO_2, is a brown gas.

Figure 1 shows a mixture of the two gases at equilibrium in a sealed container.

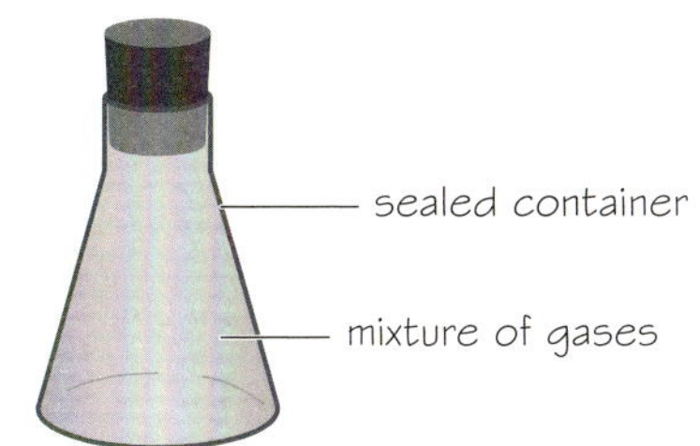

Figure 1

The equation for the reaction is:

$$N_2O_4(g) \rightleftharpoons 2NO_2(g)$$

(a) Explain what is meant by the term 'dynamic equilibrium'. **(2 marks)**

....

(b) Predict what you would observe if the pressure of the reaction vessel was increased. **(1 mark)**

....

(c) Explain why the mixture turns a deep brown colour when the reaction vessel is placed in a beaker of hot water. **(3 marks)**

....

You will already have done some work on equilibrium. Before starting the **skills boosts**, rate your confidence for each skill. Colour in the bars.

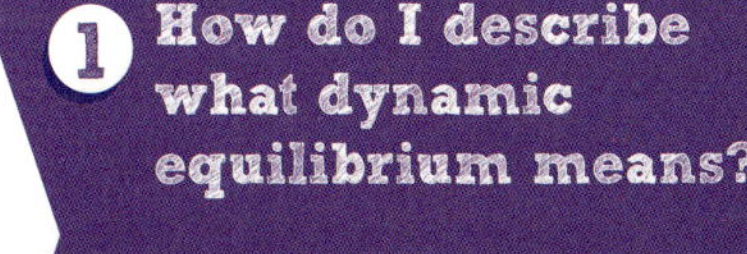

1 How do I describe what dynamic equilibrium means?

2 How do I predict changes in equilibrium position caused by temperature changes?

3 How do I predict changes in equilibrium position caused by concentration or pressure changes?

Chemistry

Get started

For equilibrium to occur there must be a reversible reaction taking place in a closed system under certain conditions of temperature, concentration and pressure.

Changes in temperature, concentration or pressure will change the position of equilibrium. Generally, the equilibrium position moves in the direction that reduces the effect of the change in the closed system.

1 A reversible reaction is a reaction in which the products can react to produce the original reactants.

For example, the decomposition of ammonium chloride:

ammonium chloride ⇌ hydrogen chloride + ammonia

a Write the forward reaction ....

b Circle the reactants of the forward reaction.

c Draw a box around the products of the forward reaction.

d Highlight the reactants of the reverse reaction.

e Underline the products of the reverse reaction.

f Complete the student's sentence below.

Look at the direction of the half arrow. The forward reaction goes from left to right.

The reactants of the forward reaction are .... as the products of the reverse reaction.

2 Identify the closed systems.

Nothing can get into or out of a closed system.

Tick **two** boxes.

A A conical flask with a tight-fitting bung. ☐

B A boiling tube with a delivery tube. ☐

C A beaker on a balance. ☐

D A bottle with a screw top. ☐

3 Draw linking lines to complete **four** sentences. One has been done for you.

A During an exothermic reaction

B During an endothermic reaction

a heat energy is taken in from the surroundings.

b the temperature increases.

c the temperature decreases.

d there is no change in heat energy.

e heat energy is given out to the surroundings.

'Exo' means 'out' as in exit and 'endo' means 'in' as in entrance, so exothermic means heat is transferred out, endothermic means heat is transferred in.

4 The diagram shows the particles in a mixture at different concentrations.

A B C

Write a letter in the box to identify the mixture with:

a the lowest concentration of ● ☐

b the highest concentration of ● ☐

How do I describe what dynamic equilibrium means?

At dynamic equilibrium both the forward and reverse reactions are constantly occurring at the same time.

1 When ammonia is made from nitrogen and hydrogen gas, the reaction reaches dynamic equilibrium.

$$N_2(g) + 3H_2(g) \rightleftharpoons 2NH_3(g)$$

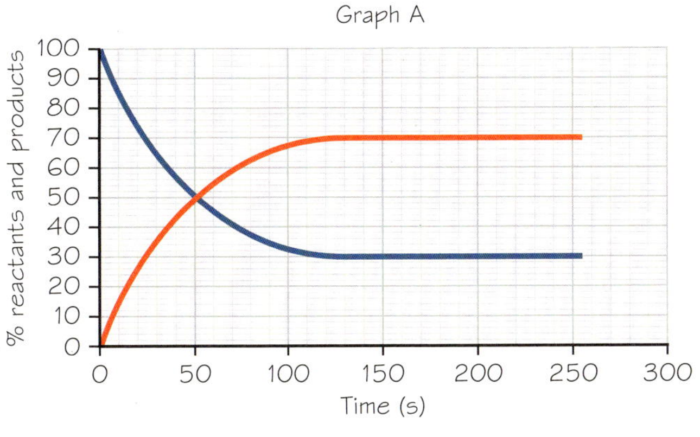

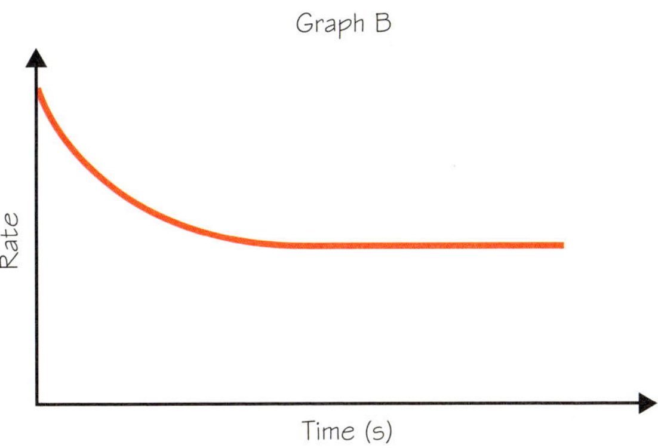

The graphs show the reaction of nitrogen and hydrogen to produce ammonia.

Graph A shows how the percentage of reactants and products of the forward reaction change over time.

Graph B shows how the rate of the forward reaction changes over time.

The rate of a reaction is the speed at which reactants are turned into products.

a One line on **graph A** shows how the amount of reactants changes over time. Label this line **R**.

Consider what happens to the reactants during the reaction.

b One line on **graph A** shows how the amount of products changes over time. Label this line **P**.

c On each line on **graph A**, write **E** at the point when dynamic equilibrium is reached.

The percentages of reactants and products no longer change.

d Draw another line on **graph B** to show how the rate of the reverse reaction changes with time.

2 Circle the correct words or phrases in **bold** to complete the sentence.

At dynamic equilibrium the amount of substances in a closed system **increases / remains the same / decreases**. This is because the rate of the forward reaction is **more than / equal to / less than** the rate of the reverse reaction.

You need to think about what is happening at the particle level in a reaction that looks as though it is staying the same.

Equilibrium means 'in balance'. Dynamic means that there is 'continual change'.

Chemistry

2 How do I predict changes in equilibrium position caused by temperature changes?

The position of equilibrium indicates the relative amounts of reactants and products found in the reaction vessel. Changing the temperature of the system moves the position of equilibrium, and therefore the relative amounts of reactants and products, in order for the reaction to reverse the change. The rule is:

> If a dynamic equilibrium is disturbed by changing the conditions, the reaction moves to counteract the change.

When the temperature is increased, the equilibrium shifts in the direction of the endothermic reaction. It transfers heat energy from the surroundings.

When the temperature is decreased, the equilibrium shifts to the direction of the exothermic reaction. It transfers heat energy to the surroundings.

1 The diagram shows a reaction that is in dynamic equilibrium.

forward reaction

$H_2(g) + I_2(g) \rightleftharpoons 2HI(g)$

reverse reaction

a The forward reaction is exothermic. The reverse reaction is endothermic.

Write the words exothermic and endothermic on the correct arrows on the diagram.

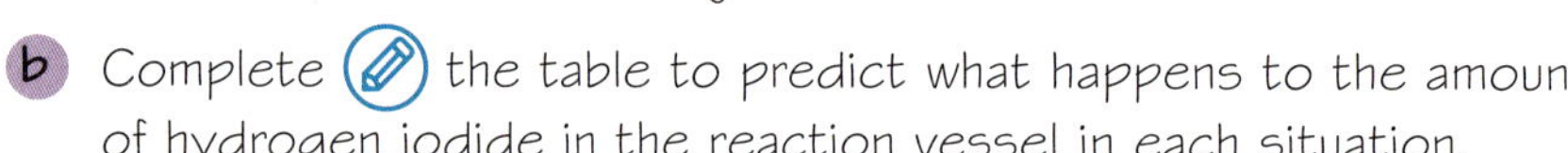

b Complete the table to predict what happens to the amount of hydrogen iodide in the reaction vessel in each situation.

Reaction situation	Amount of hydrogen iodide
At dynamic equilibrium	It stays the same
The position of equilibrium moves to the left	
The position of equilibrium moves to the right	

Look at the chemical equation. When the position of equilibrium moves it favours the reaction in the direction it is moving **or** the rate of the reaction in the direction it moves increases.

c Complete the sentences by circling the correct word in each pair in **bold**.

Decreasing the temperature moves the position of equilibrium in the direction of the **endothermic / exothermic** reaction in order to increase the temperature again.

Increasing / Decreasing the temperature moves the position of equilibrium in the direction of the endothermic reaction.

Use the above rule to predict the changes.

2 Complete the table for the reaction below, which is in dynamic equilibrium.

$$CO(g) + 2H_2(g) \rightleftharpoons CH_3OH(g)$$

Temperature change	Effect on position of equilibrium	Reason
Increase		The forward reaction is exothermic and the reaction changes to decrease the temperature.
	Moves to the right	

How do I predict changes in equilibrium position caused by concentration or pressure changes?

The position of a dynamic equilibrium changes if the concentration of a reactant or product is changed. This alters the relative amounts of reactants and products at equilibrium.

The position of equilibrium shifts to reduce the effects of any changes to the system. If the concentration of a substance is increased, the equilibrium shifts in the direction that uses up the substance that has been added. If the concentration of a substance is decreased, the equilibrium shifts in the direction that forms more of the substance that has been removed.

Changing pressure in reactions involving gases also affects the equilibrium position.

1 The diagram shows the particles of some substances in dynamic equilibrium.

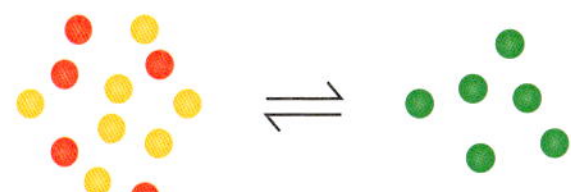

a If the concentration of ● (yellow) is increased, how will the system react to restore the equilibrium?

Tick one box.

A Produce more ● (red) ☐

B Produce more ● (yellow) ☐

C Produce more ● (green) ☐

D Shift the position of equilibrium towards the left ☐

Remember The system is closed so no other particles can enter or leave the system on their own.

Apply the rule that states: 'If a dynamic equilibrium is disturbed by changing the conditions, the reaction moves to counteract the change.'

b If the concentration of ● (green) is decreased, how will the system react to restore the equilibrium?

Tick two boxes.

A Produce less ● (red) + ● (yellow) ☐

B Produce less ● (green) ☐

C Shift the position of equilibrium towards the left ☐

D Shift the position of equilibrium towards the right ☐

When working with gases, the pressure can also change the equilibrium position.

Increasing the pressure moves the equilibrium position to the side that forms the **smallest** number of molecules of gas, as this reduces the pressure.

2 Complete these sentences about the reaction between nitrogen and hydrogen to form ammonia.

$$N_2(g) + 3H_2(g) \rightleftharpoons 2NH_3(g)$$

On the left-hand side of the equation, there are .... molecules of gas. On the right-hand side of the equation, there are .... molecules of gas. More molecules produce more pressure, so an increase in pressure will move the position of equilibrium to the .... This would result in an increase in .... molecules.

Chemistry

Get back on track

Sample response

Here are some exam-style questions. Use the student responses to these questions to improve your understanding of dynamic equilibrium.

Exam-style question

1 An equilibrium forms between chlorine and two different iodine chlorides:

$ICl(l)$	+	$Cl_2(g)$	$\rightleftharpoons$	$ICl_3(s)$
dark brown		green		yellow

(a) Describe **two** conditions needed for dynamic equilibrium to occur. **(2 marks)**

Here is student A's response.

Rate of the reactions is the same.

Describe means recall some facts. **Two** is in bold, so you must recall two facts.

Here is student B's response.

In a closed system the rate of the forward reaction is equal to the rate of the reverse reaction.

1 Give **two** reasons why student B's answer gains more marks than student A's answer.

1 ....

....

2 ....

....

Exam-style question

(b) When the equilibrium mixture is heated, the colour becomes a darker brown.

Explain whether the reverse reaction is endothermic or exothermic. **(2 marks)**

....

Here is student C's response.

Endothermic because the reverse reaction is favoured by higher temperatures.

Here is student D's response.

Endothermic because the reverse reaction lowers the temperature.

To answer this question, extract the relevant information from the question and then apply the rules about equilibrium position.

2 Explain why you think student C's answer was awarded 2 marks but student D's answer was only awarded 1 mark.

....

....

Your turn!

It is now time to use what you have learned to answer the exam-style question from page 81.

Remember to read the question thoroughly, looking for information that may help you.

Make good use of your knowledge from other areas of chemistry.

Exam-style question

1 Dinitrogen tetroxide, N_2O_4, is a colourless gas.

Nitrogen dioxide, NO_2, is a brown gas.

Figure 1 shows a mixture of the two gases at equilibrium in a sealed container.

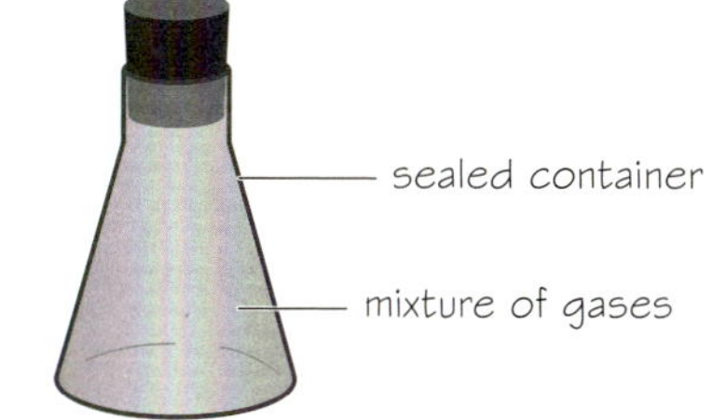

Figure 1

The equation for the reaction is:

$$N_2O_4(g) \rightleftharpoons 2NO_2(g)$$

(a) Explain what is meant by the term 'dynamic equilibrium'. **(2 marks)**

....

....

....

You need to consider both the forward and reverse reactions.

An answer to an explain question needs a reason.

(b) Predict what you would observe if the pressure of the reaction vessel was increased. **(1 mark)**

....

....

Look at the total number of molecules of gas on each side of the equation.

When asked to observe, only write down what you can see.

(c) Explain why the mixture turns a deep brown colour when the reaction vessel is placed in a beaker of hot water. **(3 marks)**

....

....

....

Use the colour change to work out which way the position of equilibrium has moved.

Need more practice?

Exam questions may ask about different parts of one topic, or parts of more than one topic. Questions about dynamic equilibrium could occur as:

- questions about that topic only
- part of a question about rates of reaction
- part of a question about industrial processes such as the manufacture of ammonia.

Have a go at these exam-style questions. 

Exam-style questions

1 Fizzy drinks are made by forcing carbon dioxide gas into the drink under high pressure.

This causes the $CO_2(g)$ to dissolve in the drink to form $CO_2(aq)$.

A dynamic equilibrium is established in an unopened bottle: $CO_2(g) \rightleftharpoons CO_2(aq)$

(a) Compare the rate of the forward and reverse reactions when the drinks bottle is closed. Suggest a reason for your answer. **(2 marks)**

....

(b) Predict what will happen to the position of equilibrium when the bottle is opened. Give a reason for your answer. **(2 marks)**

....

2 Sulfuric acid is manufactured in different stages.

In one stage, sulfur dioxide, SO_2, is converted to sulfur trioxide, SO_3, in an exothermic reaction.

The equation for the reaction is: $2SO_2(g) + O_2(g) \rightleftharpoons 2SO_3(g)$

(a) Explain why a pressure of 200 atmospheres would most favour the forward reaction. **(2 marks)**

....

(b) Explain the effect of increasing temperature on this reaction. **(2 marks)**

....

Boost your grade

You may be asked to interpret graphs and other data in a range of different contexts.

How confident do you feel about each of these **skills**? Colour in the bars.

1 How do I describe what dynamic equilibrium means?

2 How do I predict changes in equilibrium position caused by temperature changes?

3 How do I predict changes in equilibrium position caused by concentration or pressure changes?

5 Energy changes in reactions

This unit will help you understand energy changes in reactions.

In the exam, you will be asked to answer questions such as the one below.

Exam-style question

1 Methane burns completely in oxygen to form carbon dioxide and water.

$$CH_4 + 2O_2 \rightarrow CO_2 + 2H_2O$$

$$H{-}C(H)(H){-}H + 2(O{=}O) \longrightarrow O{=}C{=}O + 2(H{-}O{-}H)$$

The reaction profile for this reaction is shown in **Figure 1**.

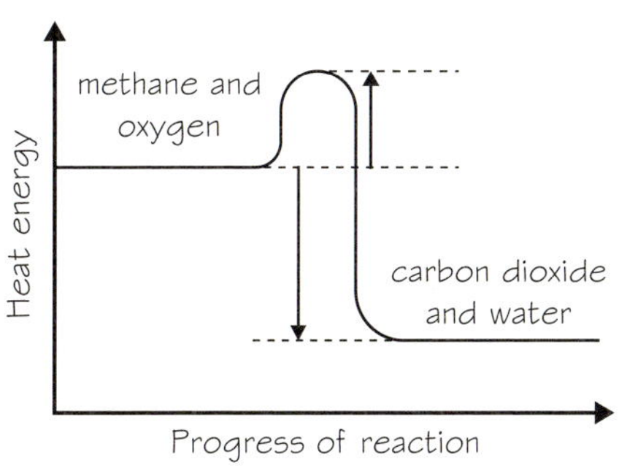

Figure 1

(a) Explain how the reaction profile shows that this reaction is exothermic. **(2 marks)**

....

The energies of some bonds are shown in **Figure 2**.

Bond	C—O	C—H	H—H	O—H	C=O
Bond energy ($kJ\,mol^{-1}$)	358	413	436	464	805

Figure 2

(b) The overall energy change for this reaction is $-818\,kJ\,mol^{-1}$. Calculate the bond energy for the bond O=O using the equation for the reaction and the information in **Figure 2**. **(4 marks)**

....

You will already have done some work on energy changes in reactions. Before starting the **skills boosts**, rate your confidence in each skill. Colour in the bars.

1 How do I interpret reaction profile diagrams?

2 How do I draw reaction profile diagrams?

3 How do I calculate energy changes using bond energies?

Get started

An exothermic reaction is one that transfers heat energy to the surroundings so the temperature of the surroundings increases.

An endothermic reaction is one that takes in heat energy from the surroundings so the temperature of the surroundings decreases.

Remember The 'surroundings' is everything other than the reacting substances themselves. This includes the water if it's a solution, the container and the air around the container.

We can work out whether the reaction is exothermic or endothermic by measuring the temperature change during a reaction.

1 Tick to show which of the reactions below are endothermic and which are exothermic.

	Endothermic	Exothermic
Hand warmers	☐	☐
Ice pack	☐	☐
Neutralisation reactions	☐	☐
Burning fuels	☐	☐
Thermal decomposition of copper carbonate	☐	☐

2 Complete the table below.

Reaction	Initial temperature (°C)	Final temperature (°C)	Temperature change (°C)	Exothermic or endothermic
1	20	25		exothermic
2	20	17		
3	21		–36	
4	19		40	

A negative temperature change means that the temperature has decreased during the reaction.

3 Fill in the gaps by using words from the box below. You can use the words once, more than once or not at all.

activation particles maximum energy physical minimum collide chemical

.... reactions can occur only when reacting particles .... with each other with sufficient energy. The .... amount of energy that particles must have to react is called the .... energy.

How do I interpret reaction profile diagrams?

A reaction profile diagram is used to model the energy change during a chemical reaction. In these diagrams, energy stored in the chemical bonds of the reactants and products is represented by horizontal lines. The higher the line, the greater the chemical energy stored in the bonds.

The diagram below shows a simple reaction profile for chemical reactions A and B.

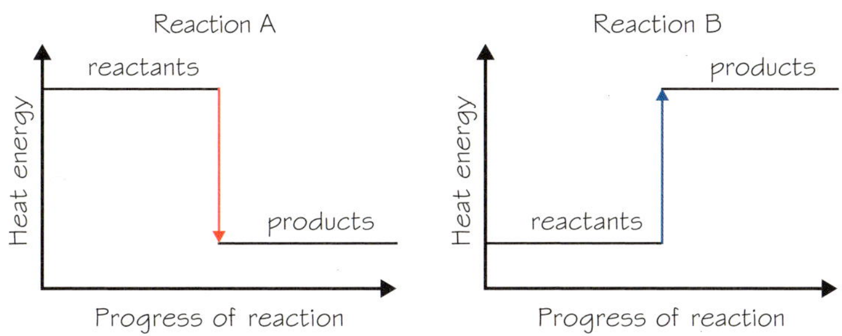

1. Identify three statements that are true.

Tick **three** boxes.

a Reaction A is endothermic. ☐

b In reaction B, the relative energy of the reactants is less than that of the products. ☐

c In reaction A, heat energy is given out to the surroundings. ☐

d In reaction B, the temperature of the surroundings will increase during the reaction. ☐

e Reaction A is exothermic. ☐

f In reaction A, the relative energy of the products is more than the relative energy of the reactants. ☐

The diagram below shows the reaction profiles for reactions C and D.

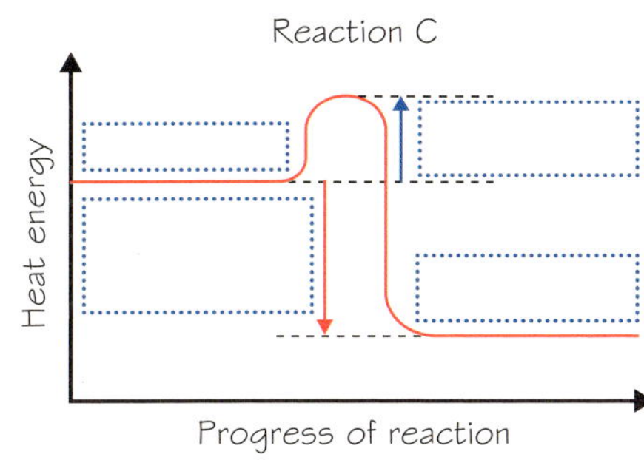

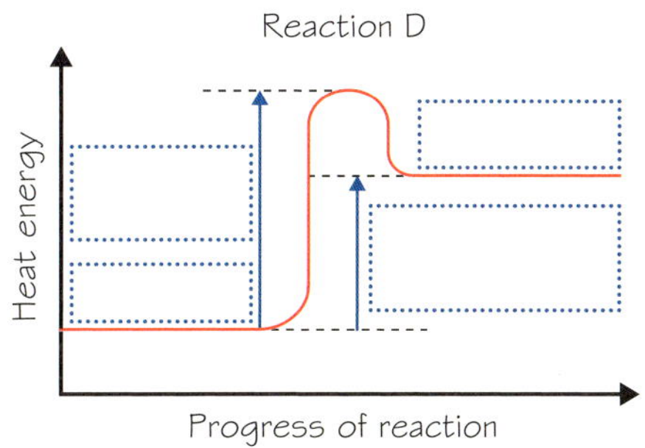

When using the term 'relative energy' it means you are comparing the amount of energy stored in different chemical bonds.

2. a Label the reaction profiles for reactions C and D using the words in the box below. You may use each word once, more than once, or not at all.

The energy needed to get the reaction going is shown by the 'hump' on the reaction profile.

reactants	products	temperature	activation energy	time	overall energy change

b Complete the sentences.

Reaction C is exothermic because energy is .... the surroundings.

Reaction D is endothermic because ....

Chemistry

2 How do I draw reaction profile diagrams?

A reaction profile diagram can be drawn successfully by following three simple steps.

1. Follow the steps below to draw a reaction profile diagram on the axes provided.

Step 1 Work out whether the reaction is exothermic or endothermic.

> The energy needed to break one mole of a particular covalent bond is the bond energy. It is measured in kilojoules per mole or kJ mol^{-1}.

a Identify which sources of information can be used to determine whether a reaction is exothermic or endothermic.

Tick **two** boxes.

A The word equation

B Temperature change data

C Colour change observations

D Balanced symbol equation

E Bond energy data

Heat energy

Progress of reaction

Step 2 Draw and label the reaction profile.

b Draw a line on the axes on the right to show the relative energy of the reactants in an exothermic reaction.

c Draw a line on the axes on the right to show the relative energy of the products in an exothermic reaction.

> **Remember** to leave a space between the two lines you draw so that you can add the activation energy later.

Step 3 The activation energy is the minimum amount of energy needed by colliding particles for a reaction to occur. The activation energy is drawn as a 'hump' between the reactant and product lines.

d Draw a 'hump' between the reactant and product lines on the graph to show the activation energy.

2. Citric acid was added to a solution of sodium hydrogen carbonate. The following were observed:

Initial temperature of sodium hydrogen carbonate solution was 20 °C.
Final temperature of the mixture after adding citric acid was 17 °C.

Complete the sentences by circling the correct word in each pair.

The reaction is **exothermic / endothermic** because the temperature **decreased / increased** during the reaction. The relative energy of the reactants is **greater / less** than the relative energy of the products.

3. The axes for a reaction profile are shown on the right.

a Label the *x*-axis and the *y*-axis.

b Draw and label a line to show the relative energy of the reactants in an endothermic reaction.

c Draw and label a line to show the relative energy of the products in an endothermic reaction.

d Draw a 'hump' between the reactant and product lines.

e Add arrows and label the following on your reaction profile:

i the activation energy

ii the overall energy change.

3 How do I calculate energy changes using bond energies?

During a chemical reaction, bonds in the reactants break and new bonds are made to form the products. Bond energies can be used to calculate the overall energy changes for a reaction.

1 Complete the sentences by circling the correct word in each pair in bold.

Breaking bonds requires energy. It is **endothermic / exothermic**.

Making new bonds gives out energy. It is **endothermic / exothermic**.

The **bond / chemical** energy is the energy needed to break a chemical bond or the energy given out when new bonds are made.

Hydrogen gas reacts with chlorine gas to produce hydrogen chloride.

$H_2(g) + Cl_2(g) \rightarrow$ .... $HCl(g)$

H—H + Cl—Cl → 2H—Cl

Bond	Bond energy in kJ / mol[1]
H—H	436
Cl—Cl	243
H—Cl	432

2 The energies of the bonds involved are listed in the table.

a Balance the equation above.

When balancing an equation make sure that the number of atoms on each side of the equation is the same.

Calculate the energy change during the reaction using the following steps.

b **Step 1:** Calculate the energy needed to break the bonds in the reactants.

Bonds broken	Energy taken in
1 × H—H	= .... $kJ\,mol^{-1}$
1 × Cl—Cl	= .... $kJ\,mol^{-1}$
Total energy in	= .... $kJ\,mol^{-1}$

When working out the number of bonds present, it is often helpful to write out the displayed formulae.

c **Step 2:** Calculate the energy given out when the bonds are made in the products.

Bonds made	Energy given out
2 × H—Cl	= 2 × .... $kJ\,mol^{-1}$
Total energy out	= .... $kJ\,mol^{-1}$

d **Step 3:** Calculate the overall energy change.

Remember the units.

Energy change = energy in – energy out

= .... – .... = ...

e Explain what the negative sign in the answer shows about the energy in and the energy out.

....

f State whether the reaction is endothermic or exothermic.

....

A negative value shows that energy is released.

Sample response

In the exam, questions about energy changes in reactions could ask you to draw or complete reaction profiles. Remember to:

- check that axes are labelled
- write on the reactant and products
- include all the bonds that are broken and made
- include the activation energy 'hump'
- check the direction of the arrows.

Here are some exam-style questions. Use the student responses to these questions to improve your answer.

Exam-style question

1 Hydrogen reacts with oxygen to produce water.

The equation for the reaction is

$$2H_2 + O_2 \rightarrow 2H_2O$$

$$2H—H + O═O \rightarrow 2H—O—H$$

Figure 1 shows the bond energies for this reaction.

Bond	Bond energy (kJ mol^{-1})
H—H	436
O═O	498
O—H	464

Figure 1

(a) Calculate the overall energy change in kJ mol^{-1} for this reaction. **(3 marks)**

(b) State whether the reaction is endothermic or exothermic. **(1 mark)**

(a) *Bonds broken = H–H and O=O so energy in = 436 + 498 = 935 kJ mol^{-1}*

Bonds made = 2 × O–H so energy out = 2 × 464 = 928 kJ mol^{-1}

Energy change = energy in – energy out = 935 – 928 = 7 kJ mol^{-1}

(b) *Endothermic*

1 The correct answer to part (a) is −485 kJ mol^{-1}. The minus sign in the answer shows that this is an exothermic reaction. Identify the two errors the student made in the calculation.

To make sure that you include all the bonds, draw out all the molecules and bonds in full and then count them up.

1 ....

2 ....

Exam-style question

2 Explain in terms of bond breaking and bond making why some reactions are endothermic. **(2 marks)**

If less energy is given out than taken in, then the reaction is endothermic.

2 Suggest a way of improving the student answer.

A good answer will include which bonds are broken and which bonds are made. It will discuss the relative energies of the bonds involved in the reaction.

....

....

Your turn!

It is now time to use what you have learned to answer the exam-style question from page 89. Remember to read the question thoroughly, looking for information that may help you. Make good use of your knowledge from other areas of chemistry.

Exam-style question

1 Methane burns completely in oxygen to form carbon dioxide and water.

$$CH_4 + 2O_2 \rightarrow CO_2 + 2H_2O$$

$$H-\underset{H}{\overset{H}{\underset{|}{\overset{|}{C}}}}-H + 2(O{=}O) \longrightarrow O{=}C{=}O + 2\left(H-O-H\right)$$

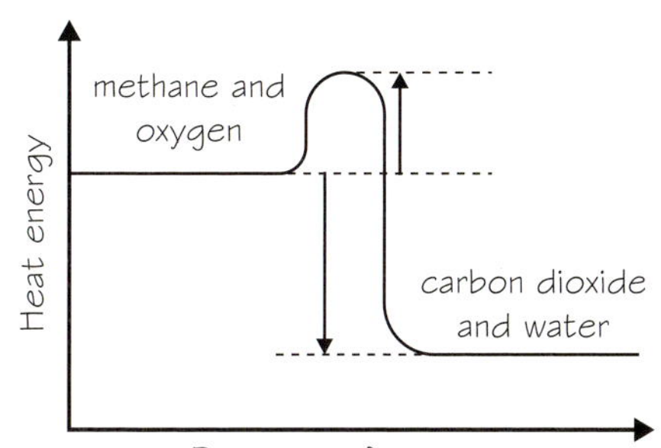

Figure 1

The reaction profile for this reaction is shown in **Figure 1**.

O=O represents one double bond and not two single bonds.

(a) Explain how the reaction profile shows that this reaction is exothermic. **(2 marks)**

....

....

....

The energies of some bonds are shown in **Figure 2**.

Bond	C—O	C—H	H—H	O—H	C=O
Bond energy ($kJ\,mol^{-1}$)	358	413	436	464	805

Figure 2

(b) The overall energy change for this reaction is $-818\,kJ\,mol^{-1}$. Calculate the bond energy for the bond O=O using the equation for the reaction and the information in **Figure 2**. **(4 marks)**

There are two molecules of oxygen in the equation for the reaction so remember to divide the final answer by 2.

....

....

....

....

....

....

....

....

....

Chemistry

Need more practice?

Exam questions often ask about different parts of a topic, or parts of more than one topic. Questions about energy changes in reactions could occur as:

- questions about that topic only
- part of a question about equilibrium or rates of reaction
- part of a question about an experiment or investigation into temperature changes.

Have a go at this exam-style question. 

Exam-style question

1 Ammonia is made from nitrogen and hydrogen in the Haber process.

The equation for the reaction is

$$N_2(g) + 3H_2(g) \rightleftharpoons 2NH_3(g)$$

$$N{\equiv}N + 3H{-}H \rightleftharpoons 2\ \text{NH}_3\ (\text{N bonded to three H})$$

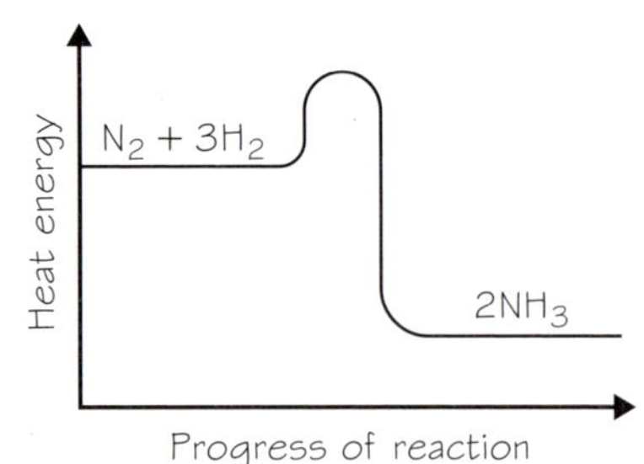

Figure 1

Look at the reaction profile shown in **Figure 1**.

(a) Describe how the energy level changes during the reaction in **Figure 1**. **(2 marks)**

....

....

....

Bond	Bond energy ($kJ\,mol^{-1}$)
H—H	436
N≡N	945
N—H	391

(b) Calculate the overall energy change in $kJ\,mol^{-1}$ for the reaction. **(4 marks)**

....

....

....

....

Boost your grade

More challenging questions may involve chemical reactions including the use of catalysts. This can make the reaction profiles, and the bond energy calculations, more demanding.

How confident do you feel about each of these **skills**? Colour in the bars.

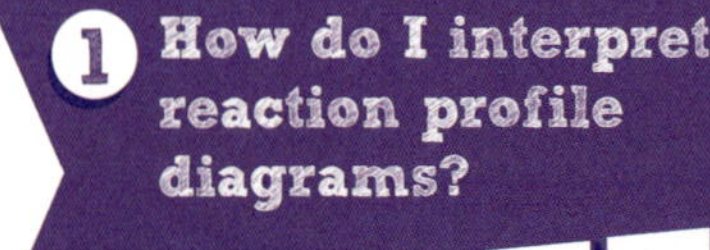

1 How do I interpret reaction profile diagrams?

2 How do I draw reaction profile diagrams?

3 How do I calculate energy changes using bond energies?

6 Answering questions about practicals

This unit will help you to answer questions based on practical work and practical situations.

In the exam, you will be asked to answer questions such as the one below.

Exam-style question

1 A student investigated the electrolysis of copper chloride solution using the apparatus shown in Figure 1.

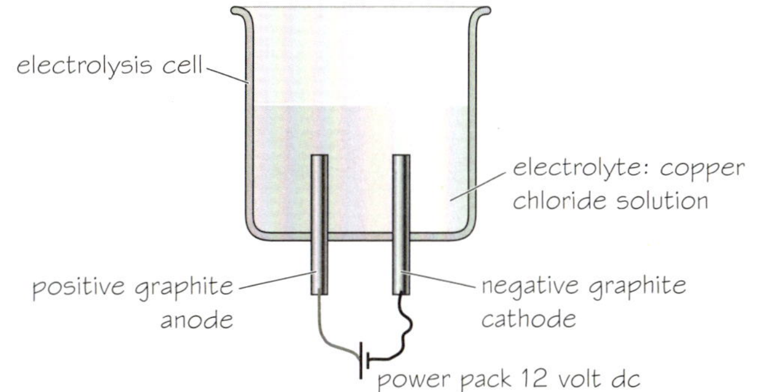

Figure 1

(a) Predict the product formed at the anode. **(1 mark)**

....

....

(b) Describe a simple laboratory test that you could carry out to test your prediction for the anode. **(2 marks)**

....

(c) The mass of copper formed at the cathode changed when the current was increased.

(i) Write a balanced half equation for the reaction at the cathode. **(1 mark)**

....

(ii) Describe **two** measurements you would make to test how the mass of copper changes when the current is increased. **(2 marks)**

....

(iii) Give **two** variables that you would control to make it a fair test. **(2 marks)**

....

(iv) Describe how more valid results could be obtained. **(2 marks)**

....

You will already have done some practical work. Before starting the **skills boosts**, rate your confidence in answering questions about practicals. Colour in the bars.

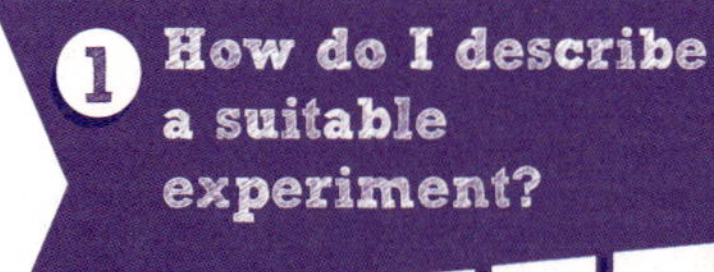

1 How do I describe a suitable experiment?

2 How do I change an investigation?

3 How do I improve the validity of results?

Chemistry

Get started

Scientists form a hypothesis, or idea, using their chemical understanding, knowledge and observations from previous experiments. They then make predictions and design suitable experiments to test the hypothesis. It is important to make sure that the results are repeatable and reproducible so that a valid conclusion can be made.

1 Draw a line to match each key word to its meaning. One has already been done for you.

Key word	Meaning
A control variable	a what you measure
B independent variable	b what you change during the investigation
C dependent variable	c what you keep constant during the investigation
D hypothesis	d the same results obtained from another person or technique
E repeatable	e based on data from an appropriate experiment
F reproducible	f an idea intended to explain certain facts or observations
G valid conclusion	g the same results obtained when the experiment is repeated

It is essential to identify the variables that might affect the results of an investigation. When designing an experiment, you must describe clearly how all these variables will be changed, measured or kept constant.

2 In each of the following experiments:

a Circle the independent variable.

b Highlight the dependent variable.

Ask yourself the questions: What is being changed? What is being measured?

A An investigation into the effect of different catalysts on the time taken to produce 10 cm^3 of oxygen during the decomposition of hydrogen peroxide.

B The effect of changing acid concentration on the temperature changes that take place during a neutralisation reaction.

C The effect of different mobile phases on the position of spots seen on a chromatogram.

The difference between the maximum and minimum values of the independent or dependent variable is known as the range. The size of the range is important when looking for patterns in the data.

3 The table shows some experimental data obtained when a student investigated how temperature affects the rate of a reaction.

Temperature (°C)	Time (s)
20	267
40	20
45	10

Complete the sentences by circling the correct word in each pair in **bold**.

A valid conclusion **can / cannot** be made from this data because the temperature range is too **small / large**.

At each temperature, there is **one data point / three data points** recorded, so it is **possible / not possible** to tell if the results are repeatable.

At least four or five data points are usually required to identify a pattern in the results.

 How do I describe a suitable experiment?

To describe a suitable experiment, you must:

- identify the independent, dependent and control variables
- use or develop a hypothesis based on your chemical knowledge or observations from previous experiments
- apply your chemical knowledge to plan an experiment to make observations and test the hypothesis.

1 A student investigated the electrolysis of aqueous solutions of salts. She made a hypothesis based on an observation from a previous electrolysis experiment.

During the electrolysis of copper sulfate solution, copper is deposited on the negative electrode. Increasing the current during the electrolysis will increase the mass of metal deposited on the negative electrode.

a Underline Ⓐ the hypothesis.

Next the student thought about the actual experiment that she needed to do to test her hypothesis. She made some notes to get started.

Make up a solution of copper sulfate or copper chloride solution to test.
Chlorine will displace iodine from a solution of iodide salts.
In electrolysis, electricity is used to split up compounds, so we need to set up an electric circuit.
Ammeters are used to measure current.
Small test tubes can be used to collect any gas given off so we can test it.
Graphite electrodes are inert.

b Highlight the notes that would help the student to design an experiment to test her hypothesis.

Make sure that you only use relevant ideas.

Finally the student set up the equipment shown in the diagram.

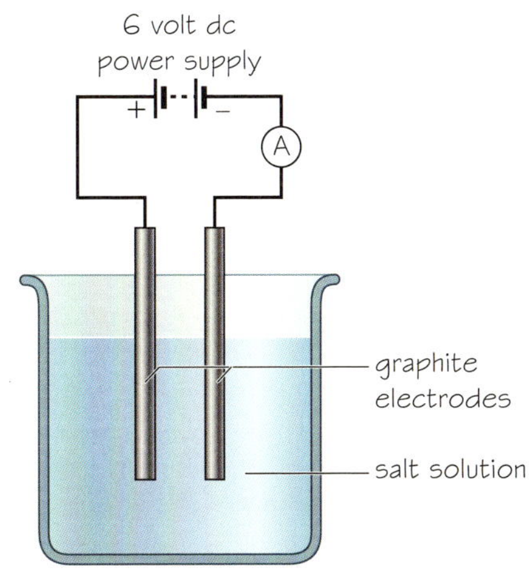

c Look at the diagram and then complete the table below. Include as many control variables as you can.

Independent variable	
Dependent variable	
Control variables	

Control variables are things that should not be changed during the experiment.

Chemistry

How do I change an investigation?

To change an investigation you need to identify the variables then change one variable and make sure that the rest are controlled. Decide on the range and collect some data.

You can widen the range in an investigation to confirm a pattern in some data. You can also carry out further tests to support or disprove your conclusion.

1 A student investigated the electrolysis of aqueous solutions of salts. His results from previous experiments are shown in the table below.

Salt solution	*Observation at the anode (+)*	*Possible product at anode*	*Observation at the cathode (-)*	*Possible product at cathode*
copper sulfate	*colourless bubbles*	*oxygen*	*brown solid on electrode*	*copper*
lead nitrate	*colourless bubbles*	*oxygen*	*grey solid on electrode*	*lead*
sodium chloride	*chlorine smell*	*chlorine*	*colourless bubbles*	*hydrogen*

Based on his chemical knowledge, the student made some predictions about his observations. He also carried out some further tests. His final conclusions are written below.

At the anode, negative ions lose electrons. So the colourless bubbles seen at the anode are oxygen gas. At the cathode, positive ions gain electrons. So the solids are copper and lead and the colourless bubbles seen at the cathode are hydrogen gas.

a The student wanted to continue the investigation to answer the following question: Is copper always produced at the cathode when a copper salt solution is electrolysed?

Below are two possible investigations (A and B) that could help answer his question.

For each investigation:

i Circle the independent variable. ii Highlight the dependent variable.

iii Write the control variables.

A The effect of changing the electrical current on the mass of copper produced.

Control variables: ....

B The effect of using different types of electrodes (graphite or copper) on the products produced at the cathode.

Control variables: ....

b The student decided to investigate the effect of changing the electrical current on the mass of copper produced at the electrode.

i What range of current should he use? Tick one box.

0.2 A ☐ 0.1–0.5 A ☐ 0.01–0.02 A ☐ 1 A ☐

ii Give a reason for your answer.

....

....

The range is the difference between maximum and minimum values of the independent and dependent variables.

3 How do I improve the validity of results?

The validity of results can be improved by ensuring that they are obtained by using a suitable procedure where variables are controlled and are accurate and precise.

Precise measurements are ones in which there is very little spread about the mean value. A result is **accurate** if it is judged to be close to a true value, repeatable and reproducible.

1 The diagram shows the results of four darts matches. Each player was trying to get as close to the centre as they could.

How would you describe each match? Write the correct letter in each box.

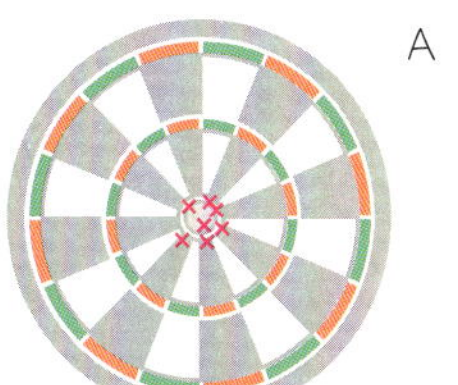

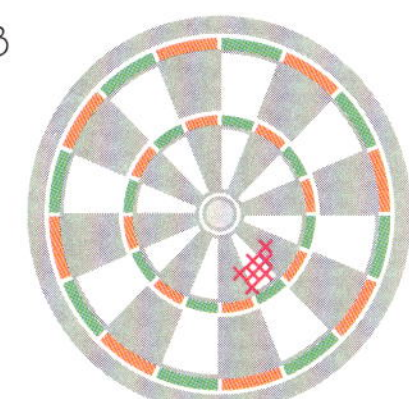
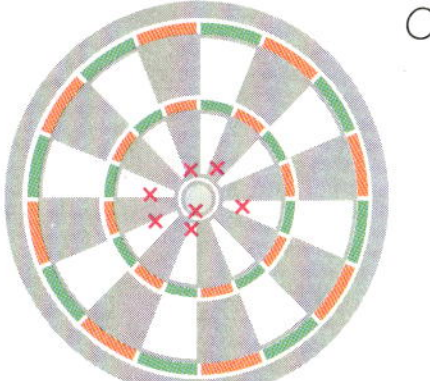
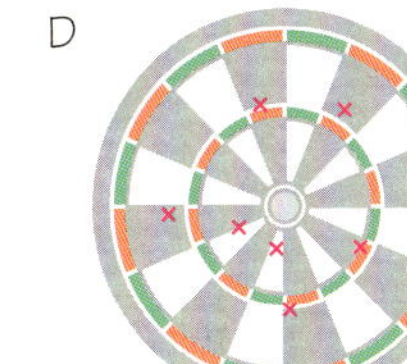

a Precise and accurate

b Not precise but accurate

c Precise but not accurate

d Not precise or accurate

2 Here are the results of copper sulfate electrolysis investigations carried out by two groups of students.

Group 1

Current (A)	0.2	0.3	0.4	0.5
Mass of cathode at start (g)	2.77	2.68	2.53	2.36
Mass of cathode at end (g)	2.85	2.79	2.67	2.99
Mass increase (g)	0.08	0.11	0.14	0.63

Group 2

Current (A)	0.2	0.3	0.4	0.5
Mass of cathode at start (g)	2.8	2.7	2.5	2.4
Mass of cathode at end (g)	2.9	2.7	2.7	2.6
Mass increase (g)	0.1	0.0	0.2	0.2

a Circle the anomalous result obtained by Group 1.

Anomalous results are odd results. They do not follow any pattern that has been observed in the results.

You can also look at Group 2's results to help you.

b Suggest a reason why the result may be anomalous.

....

There are many reasons for anomalous results, such as a variable not being controlled or writing down the wrong result.

Scientists deal with anomalous results by either repeating that part of the experiment and taking extra care, or not using that result when drawing their conclusions.

c The students in Group 1 think that they will improve the validity of the results if they repeat the experiment and calculate the mean result. Explain why they are correct.

To calculate the mean, add up all the results and divide by the total number of results.

....

d Circle the correct word in each sentence.

The data obtained by Group 2 does not show the same pattern as that obtained from Group 1. So the results are not **reproducible / repeatable**. This may be due to using a balance with a **low / high** resolution that only measures to **one decimal place / two decimal places**. Group 2 should repeat the experiment using a **precise / accurate** balance.

The resolution of a measuring instrument, such as a balance, is the smallest change in the quantity being measured, e.g. 1 g, 0.1 g or 0.01 g.

Chemistry

Sample response

Here are some exam-style questions. Use the student responses to these questions to improve your answers to questions about:

- designing experiments
- identifying and controlling variables
- improving the validity of results.

Exam-style question

1 Two students monitored the progress of a reaction by observing a colour change. Student A recorded the results as shown in Table 1 and student B recorded the results as shown in Table 2.

Mean temperature (°C)	Time taken for cross to disappear (s)
20	164
30	78
40	45
50	20

Table 1

Mean temperature (°C)	Time taken for cross to disappear (s)			Mean time taken (s)
	Trial 1	Trial 2	Trial 3	
20.1	165	175	165	165
30.2	82	83	81	82
40.0	42	42	42	42
50.1	21	20	22	21

Table 2

Suggest reasons why the results produced by student B are likely to be more valid than those produced by student A. **(3 marks)**

Student A only carried out each experiment once whereas student B did each one three times and took the average. So the data recorded by student B is repeatable. We can't tell if the results produced by student A are repeatable.

1 To improve this answer, the student also needs to consider anomalous results.

An anomalous result is one where the difference between it and the other values is much greater than the difference between the other values.

a Explain why you can't tell if any of the results produced by student A are anomalous.

....

b What advice would you give to student A?

....

c Why did student B leave data point 175 out when calculating the mean time taken at 20.1°C?

....

d Why do we get anomalous results?

....

Your turn!

It is now time to use what you have learned to answer the exam-style question from page 97. Remember to read the question thoroughly, looking for information that may help you. Make good use of your knowledge from other areas of chemistry.

Exam-style question

1 A student investigated the electrolysis of copper chloride solution using the apparatus shown in Figure 1.

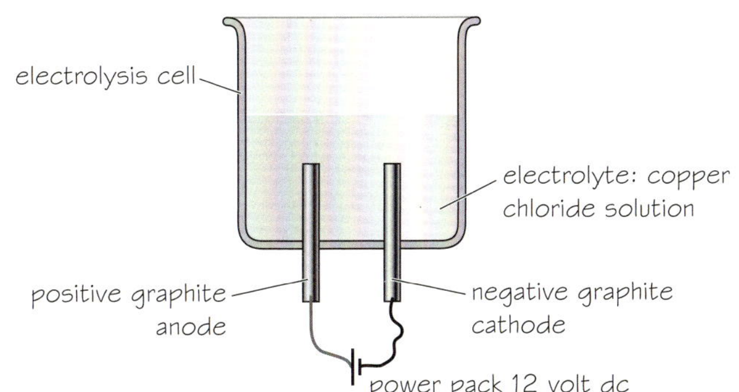

Figure 1

(a) Predict the product formed at the anode. **(1 mark)**

....

....

(b) Describe a simple laboratory test that you could carry out to test your prediction for the anode. **(2 marks)**

....

....

....

(c) The mass of copper formed at the cathode changes when the current is increased.

The half equation will show how the copper ions become copper atoms.

(i) Write a balanced half equation for the reaction at the cathode. **(1 mark)**

....

(ii) Describe **two** measurements you would make to test how the mass changes when the current is increased. **(2 marks)**

....

....

(iii) Give **two** variables that you would control to make it a fair test. **(2 marks)**

....

....

(iv) Describe how this method could be improved to produce more valid results. **(2 marks)**

....

....

Chemistry

Get back on track

Need more practice?

Exam questions may ask about different parts of a topic, or parts of more than one topic. Questions about experiments or investigations could occur as:

- questions about experiments or investigations only
- part of a question on most chemistry topics.

You may be asked to:

- describe or design an experiment
- suggest ways to improve validity of the results by collecting further evidence or extending the investigation
- evaluate the quality of the data.

Have a go at these exam-style questions. 

Exam-style questions

1 Sodium carbonate solution reacts with zinc chloride solution to produce the insoluble salt, zinc carbonate.

$$Na_2CO_3(aq) + ZnCl_2(aq) \rightarrow 2NaCl(aq) + ZnCO_3(s)$$

Describe how a pure, dry sample of zinc carbonate can be obtained. **(4 marks)**

....

....

....

....

2 Describe an experiment to produce a dry sample of potassium sulfate crystals. Write your answer on a separate sheet of paper. **(6 marks)**

Potassium hydroxide is an alkaline solution that can be used to produce a potassium salt.

Which acid produces sulfate salts?

Potassium sulfate is a soluble salt. You will need to think back to when you learnt about the preparation of soluble salts.

Word or symbol equations could be used to show your understanding of how to make a salt.

Boost your grade

Make sure you are able to use and apply your practical skills to a range of different contexts.

How confident do you feel about each of these **skills**? Colour in the bars.

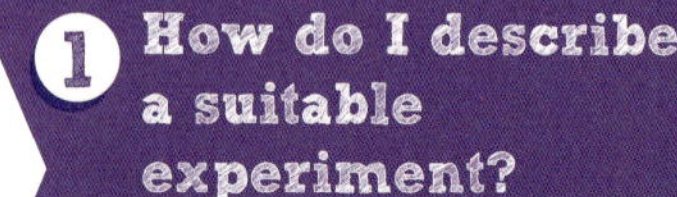

1 How do I describe a suitable experiment?

2 How do I change an investigation?

3 How do I improve the validity of results?

Get started

AO1, AO2, AO3

Answering extended response questions

This unit will help you to answer extended response questions by deciding what is being asked, and then planning a concise answer with the right detail.

In the exam, you will be asked to answer questions such as the one below.

Exam-style question

1 Describe an experiment to produce a pure, dry sample of copper chloride crystals. **(6 marks)**

....

You will already have done some work on extended response questions. Before starting the **skills boosts**, rate your confidence in each one. Colour in the bars.

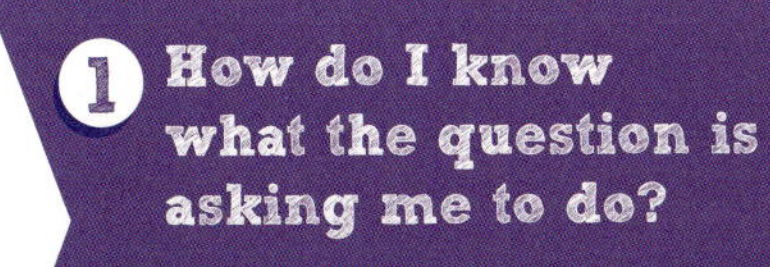

1 How do I know what the question is asking me to do?

2 How do I plan my answer?

3 How do I choose the right detail to answer the question concisely?

Chemistry

Get started

A good answer to an extended response question has the ideas well organised, correctly linked and supported with relevant scientific evidence.

1 Tick the statements that show the best way to answer an extended response question.

- [ ] Write your ideas in any order.
- [ ] Link your ideas together to show understanding.
- [ ] Write everything you know about the topic even if it isn't relevant.
- [ ] Use correct scientific vocabulary.
- [ ] Write your answer in an ordered way.

The first word in the question is usually the command word which tells you what you need to do. The Edexcel website lists the command words you need to understand.

Give, **describe**, **explain** and **predict** are common command words. **Describe** answers are likely to have more detail than **give** answers. **Explain** answers usually include the word 'because'. **Predict** means to give the expected result.

The same question can require a different response depending on which command word is used.

2 Draw lines to match each question with the two responses that best answer it.

Question

A **Give** two signs of a reaction that would be seen when a piece of sodium is dropped into a container of water.

B **Describe** what would be seen when a piece of sodium is dropped into a container of water.

C **Explain** what would be seen when a piece of sodium is dropped into a container of water.

Response

a The metal moves around on the water rapidly because the reaction is very vigorous.

b The metal moves around on the water rapidly.

c The metal disappears.

d Bubbles coming off where the sodium meets the water.

e Bubbles.

f Bubbles because hydrogen gas is given off.

How do I know what the question is asking me to do?

You can work out what the question is asking you to do by looking at the command word and anything else the question asks you to do. Think about the information that is in the question. It is important to read and think about carefully all the information provided.

The table below lists some experimental data about the reactions of metals.

Metal	Reaction with water	Reaction with hydrochloric acid
copper	no change	no change
lithium	vigorous effervescence colourless gas formed	explosive reaction
zinc	no change	bubbles of colourless gas
magnesium	a few bubbles form on surface of metal	bubbles rapidly form on surface

Exam-style question

1 Describe the relative reactivity of metals based on these experimental data. **(6 marks)**

You can follow the process below to understand what the question is asking and then answer it.

1 a Circle the command word.

b What is the command word telling you to do?

....

2 a The question asks you to describe the reactivity of metals. Highlight the other key words in the question that tell you how to do this.

'Relative' means 'compared to' in this context.

In this question, you should only use evidence in the table that is given to you, **not** other things you know.

b What are these words telling you to do?

....

....

c What information do you need to use to describe the relative reactivity?

....

Re-read the exam-style question. You will find the answer there.

You need to link the information provided to the question being asked.

3 a What are the data in the table telling you?

....

b Draw a box around each data set.

c Highlight the metals listed in the table.

d List the metals in order of how vigorously they react with:

i water: ....

ii hydrochloric acid: ....

e Explain how the reactions with hydrochloric acid help you with your final answer.

....

How do I plan my answer?

You can plan your answer by:

- thinking about the topic as a whole
- deciding which parts of the topic are relevant to the question
- structuring your answer by putting the points in a logical order.

Exam-style question

1 Describe an experiment to produce a pure, dry sample of copper sulfate crystals. Your method should be safe. **(6 marks)**

Copper sulfate is a soluble salt. You may have made a different salt in your required practical activity so you will need to apply your knowledge of chemistry.

These general questions can be used to plan practical work.

What is the chemical reaction?	How will I make it a fair test?
What reactants will I need?	What are the main steps in the method and what order should they be done in?
What equipment will I use?	What precautions must I take?
How will I use the equipment?	Do I need to use any safety equipment?
What will I observe or measure?	How will I record my results?
Consider the variables involved in the reaction. What needs to be changed or controlled?	

The question is asking you to describe the method, so you do not need to include any results or explanations.

You are making a sample of a salt and not carrying out a full investigation, so you do not need to consider variables.

1 Cross out (~~cat~~) the questions that are not relevant to making copper sulfate crystals.

Writing the word or symbol equation is a good place to start because it can be used to begin the plan.

The chemical equation for the reaction in the exam-style question is:

copper oxide + sulfuric acid → copper sulfate + water

$$CuO(s) + H_2SO_4(aq) \rightarrow CuSO_4(aq) + H_2O(l)$$

You can now plan your answer by structuring your points in a logical order as shown below and writing down some notes.

1. chemical equation	2. reactants	3. steps in method and equipment	4. equipment	5. end product	6. risk assessment

2 **a** What **three** things does the chemical equation tell you?

....

b Underline (A) the general questions above that the equations will help you to fully or partly answer.

Knowing the states of the reactants and products will help you work out what equipment is needed to measure the reactants or collect the products.

3 How do I choose the right detail to answer the question concisely?

You can answer the question concisely by:

- following your plan and making sure that you are selecting the parts of the topic that answer the question, rather than attempting to write down everything you know about the topic
- referring back to the command word to see the key elements that you should use in your answer
- expressing your ideas as clearly and precisely as possible.

Exam-style question

1 **Figure 1** shows some data about the halogens.

Halogen	Melting point (°C)	Density (kg/m^3)	Appearance	Effect of hot iron wool
fluorine	−220	1.6	pale yellow gas	bursts into flames
chlorine	−101	3.2	green gas	glows brightly
bromine	−7	3120	brown liquid	glows dull red
iodine	114	4950	purple/black solid	changes colour

Figure 1

Explain the relative reactivity of the halogens in terms of electronic configuration. **(6 marks)**

1 What is the difference between 'explain' and 'describe'? 

'Explain' means to make something clear or state the reasons for something happening.

2 Highlight the key scientific ideas that this question is asking about.

Here are some student notes for this question.

The reactivity of metals increases down the group.
The reactivity of halogens decreases down the group.
Non-metals share electrons during covalent bonding.
The electronic configuration is the way in which an atom's electrons are arranged.
The electronic configuration of chlorine is 2,8,7.
As you go down the group, the distance between the electron in the outermost shell and the nucleus increases.
Transition metals are not very reactive.
During bonding, metals lose electrons to form positive ions.
The formation of negative ions is important for reactivity.
The force of attraction depends on distance from the nucleus.

3 Highlight the notes that are relevant to the question.

Remember For some exam questions you will have to rely on your own knowledge recall as you will not be given lots of information in the question. In the exam, you will have a copy of the periodic table.

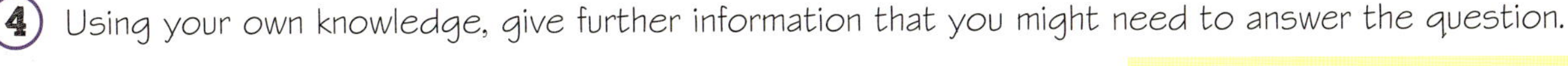

4 Using your own knowledge, give further information that you might need to answer the question. 

Include an example of the relative reactivity of the halogens. Look at the data listed in Figure 1.

Chemistry

Sample response

Use this sample student response to improve the way you answer these types of questions. Use information given in the unit to help you. Consider if the command word has been properly answered and whether all the points are covered in an organised manner.

Exam-style question

1 Describe how students could carry out an experiment to compare the reactivity of the metals copper, zinc and magnesium.

They have samples of the three metals and the metal sulfate solutions. They can use any classroom laboratory equipment. **(6 marks)**

Copper and zinc are transition metals. Magnesium is in group 2 and has two electrons on the outer electron shell. Copper is used to make electrical wires because it is a good conductor. Copper is also used to make pipes to carry water.

Put a bit of the first metal into all three of the liquids and see what it does. Repeat for the other metals. Decide which one is the most reactive and draw a results table.

Word equation:

metal + METAL sulfate ⟶ METAL + metal sulfate

This is called a displacement reaction.

1 a Circle the command word in the question.

b Cross out any irrelevant information in the student answer.

c Has the student given enough detail to answer the question? **Yes / No**

d Is the response written in an organised and logical way? **Yes / No**

Has the student included details such as: the variables involved; the experimental method; any observations needed to compare the reactivities; how they are going to decide which is the most reactive metal?

The question is asking you to describe **how** to do an experiment. You are not being asked to predict or give the results, or to explain what is happening.

You can compare the reactivity of metals by finding out which metals will displace each other. When metal 1 is added to another metal sulfate solution, if metal 1 is more reactive than the one in the sulfate solution, a new metal 1 sulfate will form and the other metal will be deposited out of the solution.

2 Now write your answer to the question on a separate sheet of paper.

- What equipment will you use and how will you use it?
- Consider the variables involved in the reaction. What needs to be changed or controlled?
- What will you observe or measure to compare the reactivity, and how will you record this?
- How will you decide the order of reactivity?
- How will you write this information in an ordered way?

Your turn!

It is now time to use what you have learned to answer the exam-style question from page 105. Remember to read the question thoroughly, looking for information that may help you. Make good use of your knowledge from other areas of chemistry.

Exam-style question

1 Describe an experiment to produce a pure, dry sample of copper chloride crystals. **(6 marks)**

1 Start by planning your answer.

a Highlight the command word in the question.

b What does the command word mean you need to do? 

....

c Which acid produces chloride salts? ....

Now think about the equipment you need and how you will use it. This is also a good place to consider safety.

> To show how you will use the equipment it is often helpful to draw some diagrams.

d Draw lines to match the process to the equipment you will use.

A Heating the reactants

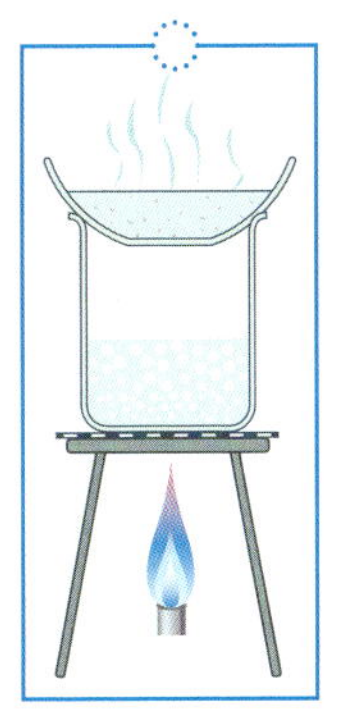

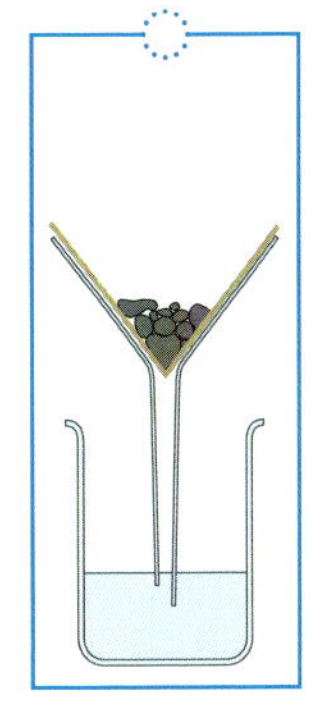

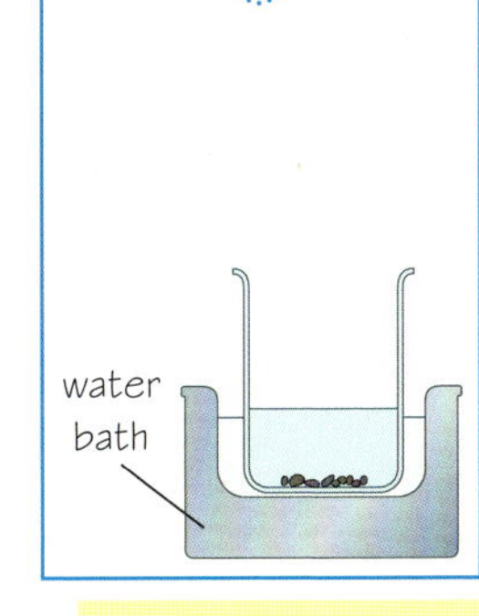

Plan your answer to get your steps in the right order.

> Copper carbonate or copper oxide can be used to produce a copper salt.

e The main steps needed to make copper chloride crystals are listed below. Write numbers in the boxes to put them into a logical order.

- ☐ Heat to evaporate off the excess water and concentrate the salt solution.
- ☐ Mix the reactants and gently warm to speed up the reaction.
- ☐ Filter off the excess unreacted copper oxide from the salt solution.
- ☐ Remove crystals and dry.
- ☐ Measure out the reactants.
- ☐ Leave to crystallise.

2 Now answer the question yourself. Use a separate sheet of paper.

> Word and/or symbol equations can be used to show your understanding of how to make a salt.

Checklist – have you:	✓
done what the command word has asked?	
organised your ideas logically?	
checked that someone could do the experiment from your instructions?	

Need more practice?

Exam questions may ask about different parts of a topic, or parts of more than one topic. Extended response questions could occur as:

- questions about any topic
- questions about an experiment or investigation.

Have a go at these exam-style questions.

Exam-style questions

1 Explain the differences in properties and uses of diamond and graphite. **(6 marks)**

2 Describe a plan to determine the effect of temperature on the reaction between copper carbonate and dilute hydrochloric acid. **(6 marks)**

Boost your grade

Make sure you are familiar with all the command words you might be asked to use. Try changing the command words of different questions and see how this changes the answers you need to give. Practise choosing which information is relevant to the question rather than just writing all you know about a topic. This will help you to be more ordered with your response.

How confident do you feel about each of these **skills**? Colour in ✎ in the bars.

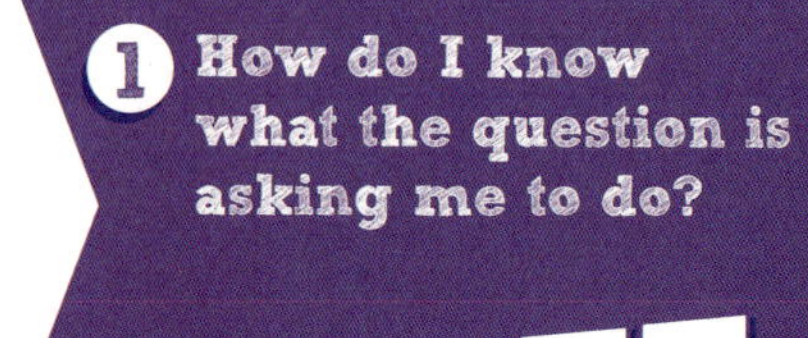

1 How do I know what the question is asking me to do?

2 How do I plan my answer?

3 How do I choose the right detail to answer the question concisely?

1 Analysing energy transfers

This unit will help you learn more about energy transfers. The unit will show you how to use an energy analysis to make predictions about what can happen and what cannot.

In the exam, you will be asked to answer questions such as the one below.

Exam-style question

1 Figure 1 shows a springy toy sitting on a desk.

When the toy is pressed down from the top, the spring inside is compressed and energy is stored.

After a while, the toy springs upwards, off the desk, due to forces produced by the spring.

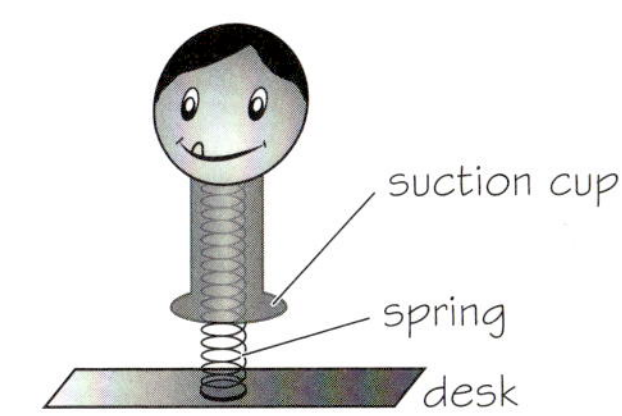

Figure 1

(a) Describe the energy transfer that takes place as the spring uncoils and the toy moves upwards to its highest point. **(3 marks)**

....

The spring constant of the spring in the toy is 50 N/m.

(b) Calculate the energy stored in the spring when it is compressed by 10 cm. **(2 marks)**

....

The mass of the toy is 60 g.

The spring in the toy has an efficiency of 0.90 when it decompresses and launches the toy upwards.

(c) Calculate the maximum height to which the toy can jump when the spring has been compressed by 10 cm. **(3 marks)**

....

You will already have done some work on energy and energy transfers. Before starting the **skills boosts**, rate your confidence for each skill. Colour in the bars.

1 How do I calculate how much energy is stored in different stores?

2 How can I use efficiency to analyse an energy transfer?

3 How can I use energy calculations to model events and make predictions?

Physics

Get started

Energy is transferred by three different processes: forces doing work, electrical currents doing work and heating effects. These processes cause one store of energy to decrease while other stores increase.

The energy decrease in one store always matches the total energy increase in other stores, so the total amount of energy is always conserved.

1 Write the process which causes energy to be transferred when:

a an object is lifted from the floor onto a shelf. ....

b an ice cube is placed into a cup of hot tea. ....

c a lamp lights up in a circuit. ....

Try not to use the term 'electricity' as it is too general. It is an electric current which transfers energy.

2 Circle the correct words to describe the energy transfer.

In a battery-operated circuit, energy is transferred to an electric motor by **an electric current / a force / heating**. The motor lifts a load through a distance and transfers energy to it by **an electric current / a force / heating**. It causes an increase in temperature by **an electric current / a force / heating**.

3 Complete this table to describe the changes in different energy stores when an energy transfer process happens.

Process	Store that decreases	Stores that increase
using a battery-powered electric motor to lift an object	chemical store of the battery	gravitational potential store of the object being lifted thermal stores of the surroundings
an object falling from a great height		kinetic store of the object falling thermal stores of the surroundings
a mechanical watch being wound up by hand	chemical store in the person winding the watch	
a kettle being used to heat water	chemical store at a power station (coal)	
an arrow being launched from a bow		

When you describe a store don't just say 'chemical store'. Make sure you say what or where that store is. For example, 'chemical store of the battery'.

4 Use the principle of conservation of energy to complete the following paragraphs.

a A student uses an electrical heater to warm up a sample of water. The heater is provided with 4000 J of energy by the electric current. The thermal store of the water increases by 3500 J while .... J is wasted by heating the surroundings.

Remember that the total energy after the transfer is equal to the total energy before the transfer.

b A lorry burns fuel with an energy content of 4.0 kJ and its kinetic store increases by 1.5 kJ. The thermal store of the truck and surroundings increases by .... kJ.

1 How do I calculate how much energy is stored in different stores?

To calculate the changes in amounts of energy in a store we take measurements of the physical properties that change. The energy stored in a spring increases when we stretch it, so we measure the change in length.

If you measure the physical properties of the store, you can calculate the amount of energy in that store. To do this you need to understand the factors that affect the sizes of the different stores.

1 Draw lines to match each energy store to the equation needed to calculate the energy in that store. One has been done for you.

Energy store	Equation
gravitational potential energy store *	$\Delta Q = m \times c \times \Delta\theta$
kinetic energy store *	$E = \frac{1}{2} \times k \times x^2$
elastic potential energy store	$Q = m \times L$
energy change during a change of state (e.g. melting)	$KE = \frac{1}{2} \times m \times v^2$
energy change when an object changes temperature	$\Delta GPE = m \times g \times \Delta h$

You need to recall the equations marked with a * because they won't be given to you on the Physics Equation Sheet.

Remember

Symbol	Stands for	Unit
m	mass	kg
h	height	m
c	specific heat capacity	J/kg°C
$\Delta\theta$	change in temperature	°C
k	spring constant	N/m
x	extension	m
L	specific latent heat	J/kg
v	speed	m/s
g	gravitational field strength	N/kg

We use different equations to calculate the amount of stored energy, depending on the energy store involved. It is important to select the right equation for the calculation. To answer questions involving energy, you will have to use the equations shown in question **1**.

2 Follow the method shown in the example to calculate the energy.

	Example	Your turn
Highlight the key data in the question	Calculate the energy change when a 4.00 kg block of aluminium with specific heat capacity of 902 J/kg°C is heated so that its temperature rises from 5.0°C to 40.0°C	Calculate the energy change when a remote-controlled toy car of mass 3.0 kg accelerates from rest (0 m/s) to 2.0 m/s
Select correct energy equation	$\Delta Q = m \times c \times \Delta\theta$	
Substitute in correct values from question	$\Delta Q = 4.00 \times 902 \times (40 - 5)$	
Calculate answer	$\Delta Q = 126\,280$	
Choose correct number of significant figures, add units	$\Delta Q = 126$ kJ	

Physics

2 How can I use efficiency to analyse an energy transfer?

The efficiency of an energy transfer tells us how much energy a useful process transfers and how much energy a process that is not useful transfers.

You need to memorise the efficiency equation:

$$\text{efficiency} = \frac{\text{useful energy transferred by the device}}{\text{total energy supplied to the device}}$$

1 A toy uses a spring to launch sponge darts through the air. The spring used to launch the darts has a spring constant of 200 N/m. When a dart is launched, the spring is compressed by 0.50 m. The efficiency of the spring launcher is 0.90. Calculate the kinetic energy of the dart.

We call the energy transferred by the process we want to happen 'useful energy'.

We call the energy transferred by a process we don't want to happen 'wasted energy'.

a Show the important data in the question.

i Circle (A) the two pieces of data that allow you to calculate the energy stored in the spring.

ii Underline (A) the efficiency of the energy transfer.

b Circle (A) the equation that allows you to calculate the energy stored in the spring.

$\Delta Q = m \times c \times \Delta\theta$	$E = \frac{1}{2} \times k \times x^2$	$Q = m \times L$
$KE = \frac{1}{2} \times m \times v^2$	$\Delta GPE = m \times g \times \Delta h$	

You do not have to memorise this equation, but you do need to select it from the Physics Equation Sheet in the exam and use it.

c Use the equation to calculate the energy stored in the spring when it is compressed.

....

....

Remember k represents the spring constant and x the extension or compression.

d Use the efficiency equation to calculate the energy transferred to the dart by the spring during a launch.

....

....

....

How can I use energy calculations to model events and make predictions?

The principle of conservation of energy states that the total amount of energy stays the same during **any** energy transfer. We use this principle to calculate the amount of change that can happen between the **start** and the **end** of any process using different energy equations.

One of the most common examples of using energy equations is analysing a fall.

Imagine an apple with mass 0.15 kg hanging from a tree at a height of 2.5 m above the ground.

We can calculate the energy stored gravitationally by the apple using an equation.

1 Name the equation you should use to calculate the gravitational potential energy of the apple.

....

2 Use this equation to calculate the energy stored. The gravitational field strength is 10 N/kg.

....

....

3 Complete the following sentence using the words **increases** and **decreases**.

As the apple falls, its gravitational potential energy .... but its kinetic energy ....

The principle of conservation of energy tells us that the kinetic energy increases by the same amount that the gravitational energy decreases.

4 Use the principle of conservation of energy to write down the amount of kinetic energy just before the apple hits the ground.

....

Now we use a second energy equation to find the speed of the apple as it hits the ground.

5 Fill in the correct equation for kinetic energy in the sentence below.

The apple is moving so we use the kinetic energy equation, which is ....

This equation can be rearranged to give $v^2 = \frac{2 \times KE}{m}$, showing us how to use the energy to find the speed.

6 Use this relationship to find v^2 using the mass of the apple and the kinetic energy.

....

....

Rearranging this equation is quite difficult. Here, it has been done for you, but it is very useful if you can do this yourself in the exam.

The answer to question 6 does not give the speed of the apple, so you need to work it out.

7 Find the speed of the apple just before it hits the ground.

....

....

Make sure that you give the correct units for speed and that your answer has a sensible number of significant figures.

We can also use other energy equations to analyse situations such as being launched from a catapult or even heating and cooling.

Sample response

Here are some exam-style questions. Use the student responses to these questions to improve your understanding of describing and calculating energy changes.

Exam-style question

1 A coconut of mass 1.0 kg falls from a height of 3.6 m.
Calculate the maximum speed that the coconut will reach before hitting the ground.
The gravitational field strength is 10 N/kg.

Gravitational potential energy $= m \times h = 1.0 \times 3.6 = 3.6$

Speed $v = \frac{2KE}{m} = \frac{2 \times 3.6}{1.0} = 7.2\ m/s$

(3 marks)

1 The student has used the wrong equation for gravitational potential energy. Write out the correct version of this equation and calculate the correct value.

....

....

2 The student has also made a mistake with the rearrangement of the equation linking speed to kinetic energy. Describe this mistake. ....

3 Use the correct version of the equation to calculate the speed.

....

....

Exam-style question

2 An electric kettle is provided with 30 000 J of energy by an electric current.
The heating element has an efficiency of 0.90 and is used to heat 0.5 kg of water.
Water has a specific heat capacity of 4200 J/kg °C.

(a) Calculate the change in temperature for the water during this heating process.

Temperature change = 14.3 °C

(3 marks)

The student has not shown their working for this answer and they have missed out a stage and so their answer is incorrect. If they had shown correct working they would have gained some of the marks, even if their final answer was incorrect.

4 Answer these questions to show the correct stages to go through and find the correct answer.

a What is the value of the energy transferred to the water, taking into account the efficiency of the transfer? ....

b What is the correct relationship between the temperature rise, specific heat capacity, mass and energy transferred? $\Delta\theta =$ ....

c What is the correct unit for temperature change? ....

d What is the correct answer (to two significant figures)? ....

Your turn!

It is now time to use what you have learned to answer the exam-style question from page 113.
Remember to read the question thoroughly, looking for information that may help you.
Make good use of your knowledge from other areas of physics.

Exam-style question

1 Figure 1 shows a springy toy sitting on a desk.

When the toy is pressed down from the top, the spring inside is compressed and energy is stored.

After a while, the toy springs upwards, off the desk, due to forces produced by the spring.

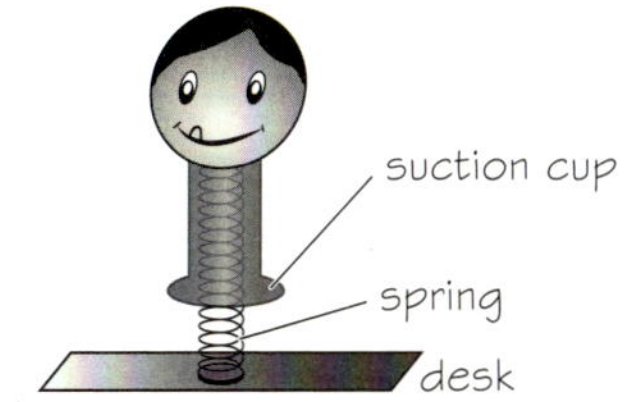

Figure 1

(a) Describe the energy transfer that takes place as the spring uncoils and the toy moves upwards to its highest point.

....

....

.... **(3 marks)**

Use the idea of energy stored by the spring, then energy stored by movement, then finally energy because an object is above the ground.

The spring constant of the spring in the toy is 50 N/m.

(b) Calculate the energy stored in the spring when it is compressed by 10 cm.

....

.... **(2 marks)**

You will need to use the equation for elastic potential energy here.

The mass of the toy is 60 g.

The spring in the toy has an efficiency of 0.90 when it decompresses and launches the toy upwards.

(c) Calculate the maximum height to which the toy can jump when the spring has been compressed by 10 cm. State your answer to two significant figures.

....

....

.... **(3 marks)**

There are two possible approaches here. Either find the height if the spring was totally efficient using the gravitational potential energy equation, and then use the efficiency equation on this value, or find 0.90 of the energy store and use this amount to calculate the height.

Need more practice?

Exam questions may ask about different parts of a topic, or parts of more than one topic. Questions about energy transfer could occur as:

- questions about that topic only
- part of a question on how objects move or have their temperature changed
- part of a question about an experiment or investigation.

Have a go at this exam-style question. 

Exam-style question

1 A student is investigating the efficiency of different bouncing balls by dropping them from a height of 2.0 m and measuring the height to which they bounce back up.

The student drops a football of mass 0.40 kg. After the first bounce, it bounces back to a height of 1.40 m.

The gravitational field strength is 10 N/kg.

(a) Calculate the gravitational potential energy of the football before it is dropped.

....

.... **(2 marks)**

(b) Calculate the maximum speed the football could be travelling at just before it hits the ground.

....

.... **(3 marks)**

(c) Describe the energy transfers during the bounce when the ball is in contact with the ground.

....

.... **(2 marks)**

(d) The efficiency of the bounce is less than 1. Calculate the efficiency of the bounce.

....

.... **(3 marks)**

Boost your grade

To boost your grade, make sure that you know the list of energy equations in the specification and how to rearrange them. You need to recall (memorise) some of the energy equations (kinetic and gravitational) but you are given the relationships for energy stored elastically or thermally. You also need to recall the efficiency equations.

How confident do you feel about each of these **skills**? Colour in the bars.

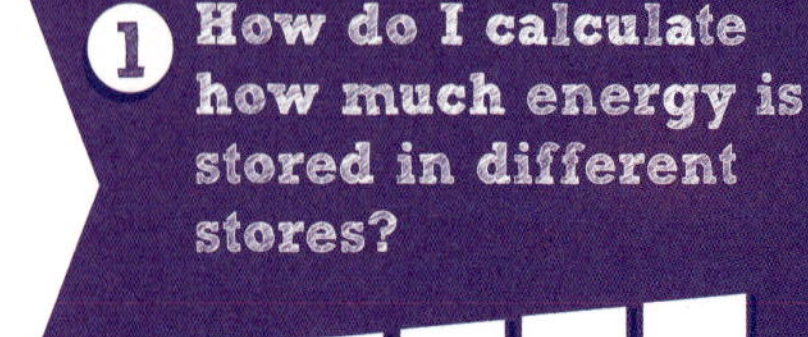

1 How do I calculate how much energy is stored in different stores?

2 How can I use efficiency to analyse an energy transfer?

3 How can I use energy calculations to model events and make predictions?

2 Newton's laws, forces and momentum

This unit will help you to learn more about how forces cause acceleration and changes to the movement of objects. These forces also cause changes to the momentum of objects during impacts.

In the exam, you will be asked to answer questions such as the one below.

Exam-style question

1 Figure 1 shows a golf ball of mass 45 g resting on a golf tee. A golfer will shortly hit it with a golf club.

(a) Give the momentum of the golf ball before the club hits it. **(1 mark)**

....

Figure 1

After the club has hit the golf ball, the ball moves away with a velocity of 80 m/s.

(b) Calculate the momentum of the golf ball immediately after it has been hit. Include the correct unit for momentum. **(2 marks)**

....

(c) Give the change in momentum for the golf club during the impact. **(1 mark)**

....

The impact between the club and the ball lasts for 0.02 s.

(d) Calculate the average acceleration of the ball during the impact. **(2 marks)**

....

(e) Use your answer to part (d) to calculate the average force acting on the ball during the impact. **(1 mark)**

....

You will already have done some work on forces and changes in velocity. Before starting the **skills boosts**, rate your confidence in each area. Colour in the bars.

1 How do I explain acceleration using Newton's laws?

2 How do I find the size of the forces causing objects to accelerate?

3 How can I describe the momentum of objects?

Physics

Get started

Resultant forces (sometimes called unbalanced forces) cause changes to the movement of objects. This change in movement is called acceleration. An object that is not accelerating can be at rest (stationary) or it can move at a constant speed in a straight line (a constant velocity).

1 There are many words used to describe motion. Draw a line to connect each key word to its correct definition. One has been done for you.

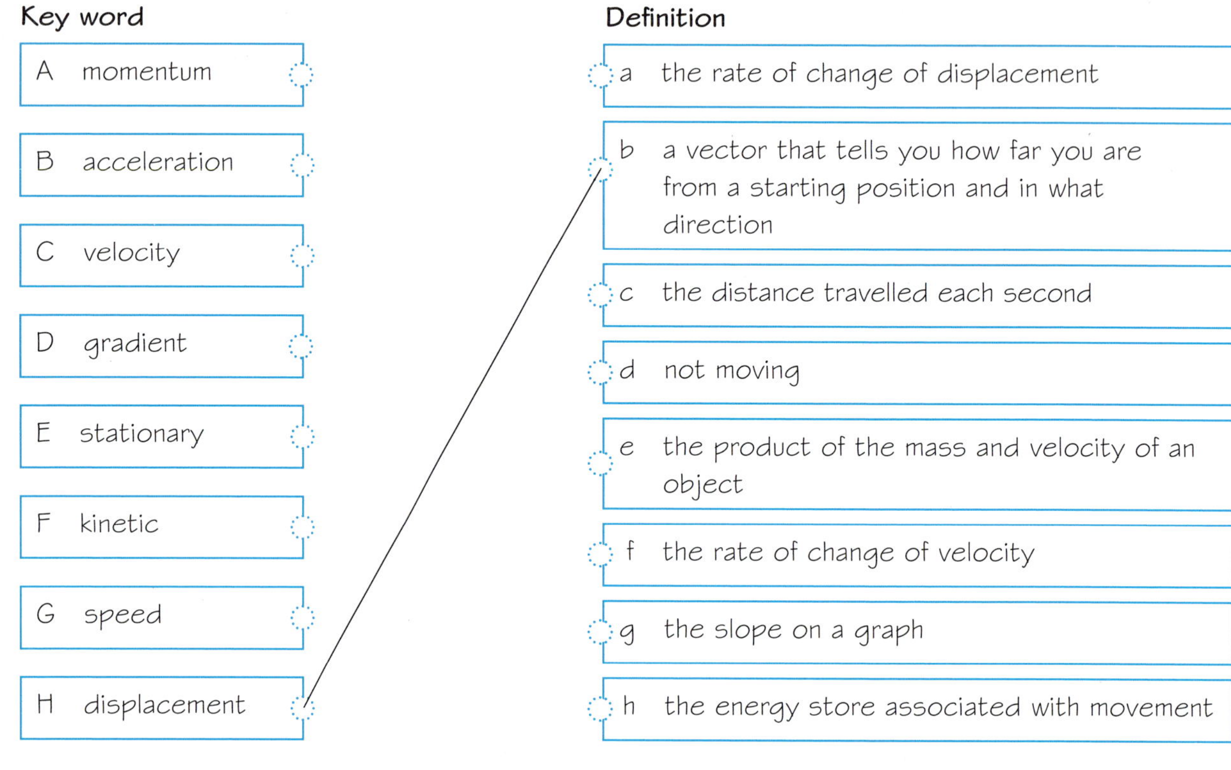

Key word
A momentum
B acceleration
C velocity
D gradient
E stationary
F kinetic
G speed
H displacement
Definition
a the rate of change of displacement
b a vector that tells you how far you are from a starting position and in what direction
c the distance travelled each second
d not moving
e the product of the mass and velocity of an object
f the rate of change of velocity
g the slope on a graph
h the energy store associated with movement

2 The diagrams below show a book at rest on a table. Label the diagrams, selecting words from the box for the labels. Each word can be used once, more than once or not at all.

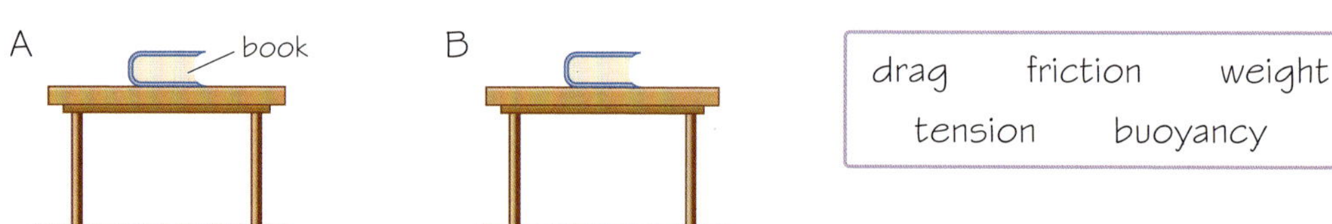

drag	friction	weight	compression
tension	buoyancy	support force	

Remember You are only drawing the forces acting on the book. The book isn't moving so there must be balanced forces acting on it.

a On diagram A, draw and label arrows to show the forces acting **only on the book**.

b On diagram B, draw and label arrows showing the forces acting **only on the table**.

Be careful with your force arrows. They should start where the force acts and point in the direction of the force. The length (*not* thickness) of the arrows should represent the size of the force.

3 The diagram on the right shows a football moving through the air after it has been kicked.

a Draw force arrows to show the forces acting on the ball as it moves through the air.

b How would these forces be different if the ball was moving faster?

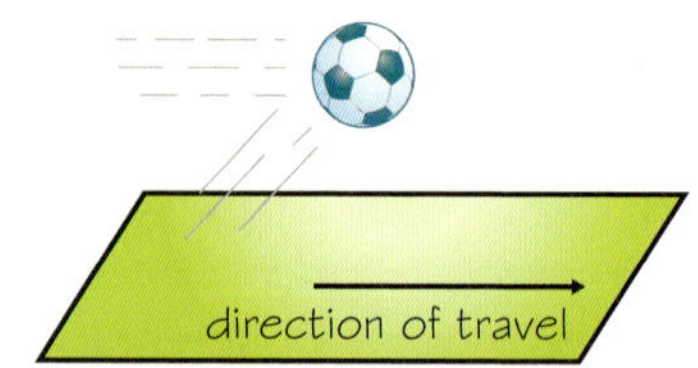

There are two forces here: a non-contact force that does not change and a contact force that acts to slow the ball down.

....

Make sure you understand what each force does to the ball as it moves. Is the force slowing the object down or speeding it up and in what direction?

1 How do I explain acceleration using Newton's laws?

Newton gave three laws of motion which describe how forces cause objects to move (or stay stationary). Newton realised that all acceleration (change in velocity) is caused by unbalanced forces acting on objects. His laws link the **resultant** force on the object to the acceleration that takes place.

1. Newton's Second Law allows us to perform calculations linking mass, acceleration and force. The relationship is usually written as $F = m \times a$.

   Draw lines to link each quantity with its symbol and unit.

Quantity	Symbol	Unit
force	a	newton, N
mass	F	metre per second squared, m/s^2
acceleration	m	kilogram, kg

2. The diagrams in the table below show the forces acting on an object.

   a Draw an arrow to represent the resultant force acting on each object. Write the size of the force. One arrow has been drawn for you.

   It is important to remember that it is the resultant force which causes acceleration.

   To find a resultant force add all the forces acting in one direction and subtract the forces acting in the opposite direction.

   b Calculate the acceleration that each object would be experiencing. Write your calculation in the table.

   To do this you could rearrange equation $F = m \times a$. This equation can be rearranged to $a = \frac{F}{m}$ and $m = \frac{F}{a}$.

	A	B	C
Object	3 kg; 5 N (left); 5 N (right); 6 N (down)	0.2 kg; 9 N (right); 7 N (left)	90 N (up); 90 N (left); 90 N (right); 6 kg; 210 N (down)
Resultant force	↓	....	....
Acceleration	$a = \frac{F}{m}$ $= \frac{\ldots\ldots}{3.0}$ $a =$ .... m/s^2	$a = \frac{F}{m}$ $= \frac{\ldots\ldots}{\ldots\ldots}$ $a =$	

How do I find the size of the forces causing objects to accelerate?

It is important to be able to find the sizes of the forces causing acceleration, particularly for vehicles. For this you need to be able to use two relationships, one after the other, as described here.

Velocity is a vector quantity. This means that it has a size and a direction. When we are finding changes in velocity we need to take into account the direction the object is moving at the start and at the end. To do this we can follow a rule like this one:

- Take movement to the right to be a positive velocity and movement to the left to be a negative velocity. For example, 6 m/s to the right is +6 m/s while 6 m/s to the left is –6 m/s.

We can do similar things with other opposite directions, such as taking movement to the north as positive velocity and movement to the south as negative velocity.

1 Complete this table to calculate the changes in velocity being described.

Use the rule that velocity to the right is positive and velocity to the left is negative.

Description of change in velocity	Identifying velocities	Calculation of change in velocity
4 m/s to the right then 8 m/s to the right	start velocity (u) = +4 end velocity (v) = +8	change in velocity = $v - u$ change in velocity = 8 − 4 change in velocity = ...
9 m/s to the right then 3 m/s to the left	start velocity (u) = +9 end velocity (v) = ...	change in velocity = $v - u$ change in velocity = −3 .... change in velocity = ...

To find the forces involved when a vehicle changes velocity we need to use $F = m \times a$. Often the acceleration, a, is not provided, so you need to go through several stages to reach the answer.

2 Follow these stages to complete the table below. The first example has been done for you.

Stage	Example	Your try
Scenario **a** Underline the information you need to calculate the acceleration.	A motorcycle has a mass of 500 kg. Calculate the force needed to accelerate it from <u>0 m/s</u> <u>to 10 m/s in 5.0 s</u>.	A car has a mass of 1200 kg. Calculate the force needed to accelerate it from 0 m/s to 5.0 m/s in 5.0 s.
b Find the change in velocity	0 to 10 m/s change in velocity = 10 – 0 = 10 m/s	0 to 5.0 m/s ....
c Find the acceleration $a = \frac{\text{change in velocity}}{t}$	$a = \frac{\text{change in velocity}}{t} = \frac{10}{5.0}$ a = 2.0 m/s	$a = \frac{\text{change in velocity}}{t} = \frac{\text{.....}}{\text{.....}}$ a = ...
d Find the size of the forces involved $F = m \times a$	$F = m \times a$ F = 500 × 2.0 F = 1000 N	$F = m \times a$ ....

Skills boost

3 How can I describe the momentum of objects?

The momentum of an object is a measurement of its 'movement'. Large fast-moving objects have high momentum while small slow-moving objects have low momentum. Momentum is calculated using the expression:

momentum (kg m/s) = mass (kg) × velocity (m/s)

In symbols, this can be written as $p = m \times v$.

Momentum is a vector quantity. This means that it has both magnitude (size) and direction.

1 The diagram below shows five balls, each with the same mass, moving in a variety of directions.

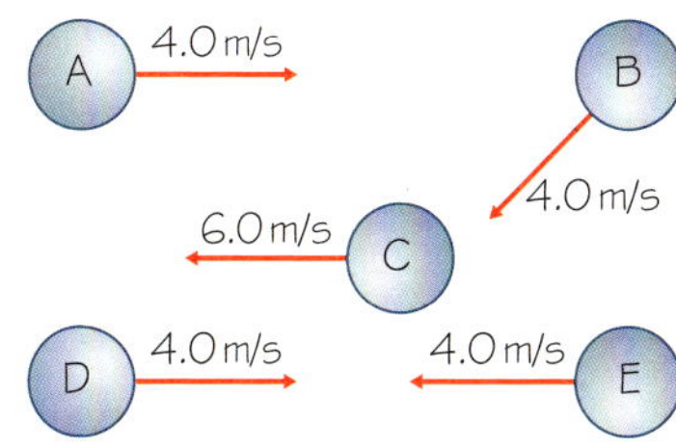

a Which balls have the same speed? ....

b Which two balls have the same velocity? ....

c Which balls have the same momentum? ....

To find the magnitude of the momentum we need to use the relationship $p = m \times v$. We also need to remember to describe the direction.

2 Calculate the momentum, $p = m \times v$, of the objects below. The first one has been done for you.

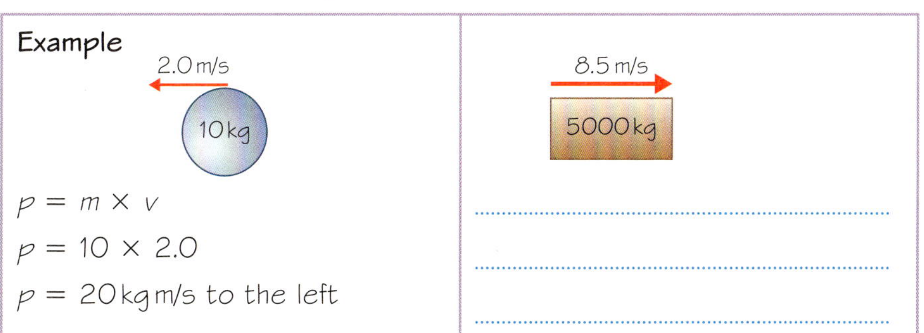

Example

2.0 m/s

10 kg

$p = m \times v$

$p = 10 \times 2.0$

$p = 20$ kg m/s to the left

8.5 m/s

5000 kg

....

....

....

Remember to give the direction of the momentum by saying whether the object is moving left or right.

The total momentum of a **system** is the sum of the momentum of each individual object taking into account the direction.

3 Two systems with momentum are shown below. For each system calculate the momentum of each object by adding the momentums together. One has been done for you.

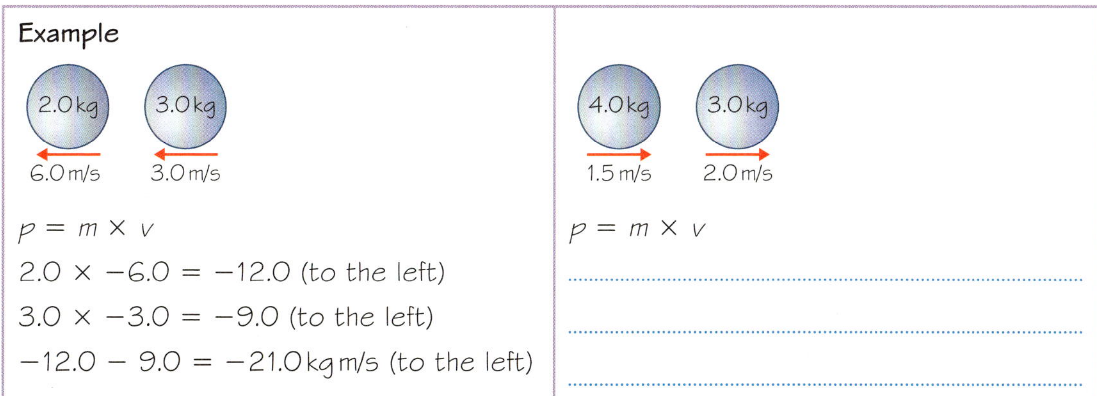

Example

2.0 kg, 6.0 m/s

3.0 kg, 3.0 m/s

$p = m \times v$

$2.0 \times -6.0 = -12.0$ (to the left)

$3.0 \times -3.0 = -9.0$ (to the left)

$-12.0 - 9.0 = -21.0$ kg m/s (to the left)

4.0 kg, 1.5 m/s

3.0 kg, 2.0 m/s

$p = m \times v$

....

....

....

Remember Take movement to the right to be a positive velocity and movement to the left to be a negative velocity. For example, 6 m/s to the right is +6 m/s while 6 m/s to the left is −6 m/s.

Physics

Sample response

Here are some exam-style questions. Use the student answers to these questions to improve your understanding of how forces cause acceleration and changes in momentum.

Exam-style question

1 A snooker ball has a momentum of 0.30 kg m/s and is travelling at 2.50 m/s.

(a) Calculate the mass of the ball. **(3 marks)**

(a) $m = \frac{v}{p} = \frac{2.50}{0.30} = 8.3333333\ kg$

(b) The ball hits the cushion of the table and bounces off in the opposite direction with a velocity of 2.0 m/s. Calculate the change in momentum for the ball. **(2 marks)**

(b) Change in momentum = change in velocity × mass = 0.5 × 8.333 = 4.17 kg m/s

1 **a** Look at the student's answer to part (a). They have used an incorrect version of the equation linking momentum, mass and velocity. Write the correct version of this equation.

Remember the definition of momentum is from momentum = mass × velocity

....

b Calculate the correct value.

....

c What other mistake has the student made in part (a) so they would not gain full marks? 

....

....

d What has the student done correctly in part (b)? 

....

e What mistake have they made in part (b)? 

Remember Momentum is a vector quantity so take a close look at the velocities.

....

....

....

....

....

f What is the correct answer to part (b)? 

....

....

Your turn!

It is now time to use what you have learned to answer the exam-style question from page 121. Remember to read the question thoroughly, looking for information that may help you. Make good use of your knowledge from other areas of physics.

Exam-style question

1 Figure 1 shows a golf ball of mass 45 g resting on a golf tee. A golfer will shortly hit it with a golf club.

Figure 1

(a) Give the momentum of the golf ball before the club hits it. **(1 mark)**

....

After the club has hit the golf ball, the ball moves away with a velocity of 80 m/s.

(b) Calculate the momentum of the golf ball immediately after it has been hit. Include the correct unit for momentum. **(2 marks)**

....

....

(c) Give the change in momentum for the golf club during the impact. **(1 mark)**

If you make a mistake calculating the momentum in part (b) this won't affect your marks for your answer to part (c) as you have already been penalised. This is known as 'carrying the error forwards'.

....

The impact between the club and the ball lasts for 0.02 s.

(d) Calculate the average acceleration of the ball during the impact. **(2 marks)**

Use the start and end velocities and the time of impact. Don't forget to give the correct unit for acceleration.

....

....

(e) Use your answer to part (d) to calculate the average force acting on the ball during the impact. **(1 mark)**

Use Newton's Second Law here.

....

....

Physics

Need more practice?

Exam questions may ask about different parts of a topic, or parts of more than one topic. Questions about momentum could occur as:

- questions about that topic only
- part of a question on how objects move or how forces change the velocity or momentum of objects
- part of a question about an experiment or investigation.

Have a go at this exam-style question. 

Exam-style question

1 A cyclist in a race is travelling at 11.0 m/s along a straight section of road. The cyclist sees a crash ahead and brakes suddenly to a speed of 5.0 m/s in a time of 1.5 s.

(a) Calculate the average acceleration of the cyclist during braking. **(2 marks)**

Look back at the equation linking acceleration, change in velocity and time.

....

....

The total mass of the cyclist and the bicycle is 60.0 kg.

(b) Calculate the average braking force the two tyres exerted on the road during the braking. **(1 mark)**

This is where you should use Newton's Second Law.

....

....

(c) Give the size of the force the road exerts on the tyres during the braking. **(1 mark)**

This question involves Newton's Third Law.

....

The cyclist is unable to stop in time and crashes at a speed of 5.0 m/s.

(d) Calculate the momentum of the cyclist and bicycle just before they crash. **(1 mark)**

....

....

Boost your grade

To boost your grade, make sure that you know how to break down calculations into stages. Acceleration is the factor that links movement and forces so practise using all the equations that involve it.

How confident do you feel about each of these **skills**? Colour in the bars.

1 How do I explain acceleration using Newton's laws?

2 How do I find the size of the forces causing objects to accelerate?

3 How can I describe the momentum of objects?

Get started

AO1

3 Radioactive decay

This unit will help you to learn more about the changes in the structure of atomic nuclei that happen during radioactive decay. It will also help you to learn about the patterns that occur during the random decay.

In the exam, you will be asked to answer questions such as the one below.

Exam-style question

1 A research scientist measured the count rate produced by a radiation detector for two different radioactive isotopes of the same element over a period of time.

The results are shown in Figure 1.

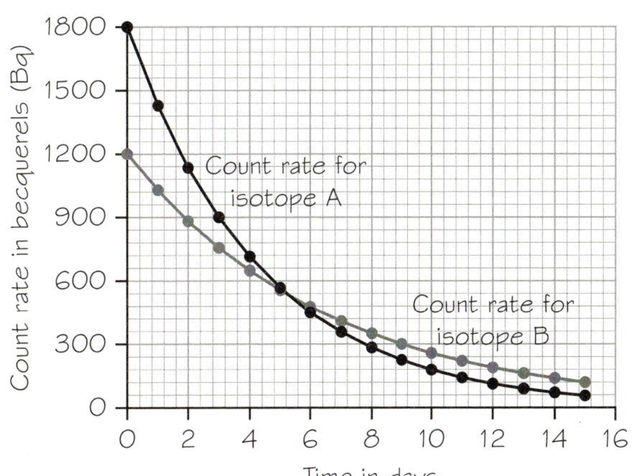

Figure 1

(a) Explain how an element can have different isotopes. **(2 marks)**

....

....

(b) State which isotope was most active at the start of the experiment. **(1 mark)**

....

(c) State which isotope had the longest half-life. **(1 mark)**

....

(d) Use the graph to determine the half-life of isotope B. **(1 mark)**

....

A sample of a different radioactive isotope (iodine-53) is known to decay by β^- (beta minus) particle emission.

(e) Complete the decay equation to show the three missing values.

$$^{131}_{53}\text{I} \rightarrow {}^{\dots}_{\dots}\text{Xe} + {}^{0}_{\dots}\text{e}$$

(3 marks)

You will already have done some work on atomic structure, radioactivity and radioactive decay. Before starting the **skills boosts**, rate your confidence in each area. Colour in the bars.

1 How do I describe the changes that happen during nuclear decay?

2 How do I write decay equations?

3 How can I use a half-life graph to analyse radioactive decay?

Physics

Get started

Atoms are the building blocks of all molecules and materials. Each atom is built from only three types of subatomic particles, **protons**, **neutrons** and **electrons**, in different numbers and arrangements.

1 Circle Ⓐ the correct answers in the table to leave the correct charges and locations of protons, neutrons and electrons.

Component	Electric charge	Location
proton	positive / negative / neutral	in the nucleus / in orbit around the nucleus
neutron	positive / negative / neutral	in the nucleus / in orbit around the nucleus
electron	positive / negative / neutral	in the nucleus / in orbit around the nucleus

Nuclear notation is used to describe a nucleus. This notation contains the element symbol, the atomic number (proton number) and the mass number (nucleon number):

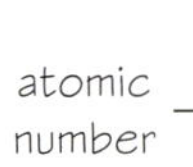

2 Complete these statements about the use of nuclear notation. Use words from the box. Each word can be used once or more than once.

atomic number | mass number | protons | neutrons | electrons | nucleus

a The atomic number is the number of .... in the nucleus.

b The mass number is the total number of .... and .... in the nucleus.

c The number of neutrons is equal to the .... minus the ....

d In an atom, the number of .... is equal to the number of protons in the

....

All carbon atoms have 6 protons but the number of neutrons can vary. We say that there are different **isotopes** of carbon. There are also different isotopes of every other element.

3 Here are three different isotopes of chlorine: $^{35}_{17}Cl$ $^{36}_{17}Cl$ $^{37}_{17}Cl$

a Write a sentence to describe what is **the same** for these isotopes. Answer in terms of subatomic particles.

....

b Write a sentence to describe what is **different** about these isotopes. Answer in terms of subatomic particles.

Try to use the terms protons and neutrons in your descriptions.

....

....

4 Use the information in question 2 to complete this table for some isotopes of atoms.

Isotope and chemical symbol	Protons (from atomic number)	Neutrons (mass number – atomic number)	Electrons (same as protons)	Atomic notation
carbon-14, C	6	14 – 6 = 8	6	$^{14}_{6}C$
carbon-12, C	6	6		
uranium-238, U		146		
				$^{17}_{8}O$

Skills boost

How do I describe the changes that happen during nuclear decay?

Some atomic nuclei are unstable. They decay to form more stable nuclei over time. When unstable nuclei decay, they emit radiation in the form of alpha particles, β^- (beta minus) particles, β^+ particles (positrons), gamma rays or neutrons.

1 Each alpha particle consists of two protons and two neutrons joined together. As this is the same as a helium nucleus, the symbol 'He' is used.

a Which of these subatomic particles in the alpha particle has a charge? ....

b How many protons are in an alpha particle? ....

c Complete this symbol for an alpha particle using nuclear notation: $^{\dots\dots}_{\dots\dots}\mathrm{He}$

> Find the atomic number for an alpha particle from the number of protons and the mass number from the number of particles.

2 Each β^- (beta minus) particle consists of a high-speed electron ejected from the nucleus as a neutron decays into a proton.

a What type of electric charge do electrons carry? ....

b The mass of an electron is so small compared with a proton that it is written as 0. The atomic number is written as −1. Use this information to complete the symbol for a β^- particle.

$^{\dots\dots}_{\dots\dots}e$

> The total number of protons and neutrons stays the same when a nucleus emits a β^- particle, but it has one less proton than before. So, to balance equations, the mass number is written as 0 and the atomic number as −1.

3 Each β^+ particle (positron) consists of a high-speed positron ejected from the nucleus as a proton decays into a neutron.

a What type of electric charge do positrons carry? ....

b Complete the symbol for a positron. $^{\dots\dots}_{\dots\dots}e$

> The total number of protons and neutrons stays the same when a nucleus emits a β^+ particle, but it has one more proton than before. So, to balance equations, the mass number is written as 0 and the atomic number as +1.

4 Complete this table to show the properties of the four most common types of radiation.

Radiation	α (alpha)	β^- (beta minus)	β^+ (positron)	γ (gamma)
Symbol for radiation in equations	${}^4_2\mathrm{He}$			
Radiation consists of	two protons and	a fast-moving .... ejected from the	a fast-moving .... ejected from the	.... emitted from the nucleus

Physics

2 How do I write decay equations?

A **decay equation** shows the changes to the nucleus, and the type of radiation emitted, during radioactive decay. For example, a radon nucleus decays to form a polonium nucleus when it emits an alpha particle. The process is shown by this decay equation:

$${}^{219}_{86}\text{Rn} \rightarrow {}^{215}_{84}\text{Po} + {}^{4}_{2}\text{He}$$

The symbol → used in the equation means 'decays into'.

In all radioactive decays, mass number and charge are 'conserved' – they do not change overall. The total mass number and total charge are the same on each side of the arrow.

1. What is the symbol for an alpha particle? ....

2. Circle Ⓐ the mass numbers for radon and the alpha particle in the equation above.

Look at the top number on each symbol in the nuclear equation.

The total mass number on the left has to be the same as the total mass number on the right.

3. Show how the equation shows this. .... = 215 + ...

The bottom number for a nucleus in a nuclear equation shows the atomic number or number of protons. This is also the charge on the nucleus, because each proton has a charge of +1. It is also the charge on the particle being emitted.

The total charge on the left has to be the same as the total charge on the right.

4. Show how the equation shows this. ....

5. Complete these decay equations.

${}^{222}_{88}\text{Ra} \rightarrow {}^{\dots}_{86}\text{Rn} + {}^{4}_{2}\text{He}$	${}^{208}_{84}\text{Po} \rightarrow {}^{204}_{\dots}\text{Pb} + {}^{4}_{2}\text{He}$

Make sure the top and bottom numbers balance on the left and right.

The β^- (beta minus) particle consists of a high-speed electron ejected from the nucleus. It is formed when a neutron decays into a proton. When a β^- particle is ejected:

- the mass number stays the same
- the proton number increases by 1.

For example, a silver nucleus decays to form a mercury nucleus by emitting a β^- particle:

$${}^{201}_{79}\text{Au} \rightarrow {}^{201}_{80}\text{Hg} + {}^{0}_{-1}\text{e}$$

6. Show how charge is conserved in the nuclear equation. 

   79 = ...

Look at the bottom number on each symbol in the nuclear equation.

7. Complete these decay equations.

${}^{6}_{2}\text{He} \rightarrow {}^{6}_{3}\text{Li} + {}^{\dots}_{-1}\text{e}$	${}^{14}_{6}\text{C} \rightarrow {}^{14}_{\dots}\text{N} + {}^{0}_{-1}\text{e}$

You don't need to know the symbols for chemical elements. They will be given in the exam.

8. Use what you have learned to complete the following table showing decay equations

Decay type	Equation	Decay type	Equation
alpha	${}^{185}_{79}\text{Au} \rightarrow {}^{\dots}_{\dots}\text{Ir} + {}^{4}_{2}\text{He}$	....	${}^{\dots}_{\dots}\text{Pa} \rightarrow {}^{227}_{89}\text{Ac} + {}^{4}_{2}\text{He}$
....	${}^{14}_{6}\text{C} \rightarrow {}^{14}_{7}\text{N} + {}^{\dots}_{\dots}\dots$	beta	${}^{8}_{3}\text{Li} \rightarrow {}^{\dots}_{\dots}\dots + {}^{0}_{-1}\text{e}$

3 How can I use a half-life graph to analyse radioactive decay?

Radioactive decay is a **random** process. We cannot predict when a particular nucleus will decay. However, when there is a large number of nuclei, we can predict a **pattern** to the decay and work out how many nuclei will be left after a certain time.

The **activity** of a sample is the number of decays that happen per second. Activity is measured in a unit called the becquerel (Bq). This activity falls in a specific pattern giving a **decay curve** shape.

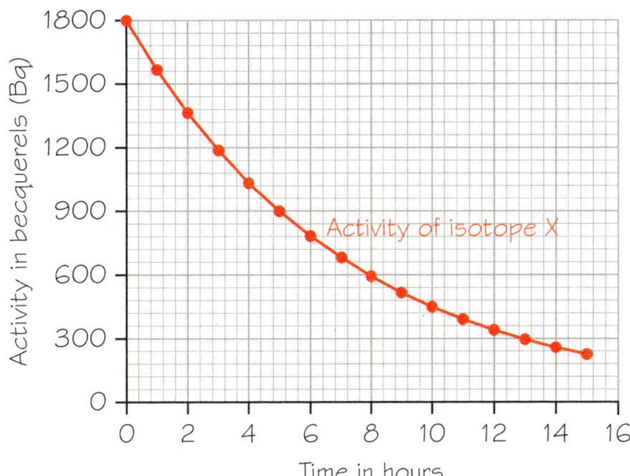

1 The graph shows how the activity of isotope X changes over time.

a What is the initial activity of the sample? Tick ✓ the correct box.

> Draw lines on the graph to help you read the values accurately.

☐ 180 Bq ☐ 1200 Bq ☐ 1800 Bq ☐ 2000 Bq

b How long does it take for the activity of the sample to halve? Tick ✓ the correct box.

☐ 0 hours ☐ 2.5 hours ☐ 5 hours ☐ 10 hours

> Use the graph to find the time taken for the activity to fall to half of your answer in part a.

c How much longer does it take for the activity to halve again (1/4 of the original activity)? Tick ✓ the correct box.

☐ 0 hours ☐ 2.5 hours ☐ 5 hours ☐ 10 hours

2 We can use the symbol → to represent one half-life passing.

What fraction of the original activity will there be after a total time of 20 hours?  ....

> 1 →(5 hours) $\frac{1}{2}$ →(5 hours) $\frac{1}{2}$ of $\frac{1}{2}$ $= \frac{1}{4}$ →(5 hours) $\frac{1}{2}$ of $\frac{1}{4}$ $= \frac{1}{8}$ →(5 hours) $\frac{1}{2}$ of $\frac{1}{8}$ $= \frac{1}{16}$
>
> Look for the pattern in the decay. Don't try to extend the graph.

The **half-life** of a radioactive sample is the time it takes for the activity of that sample to fall to half of its original value. The activity (and so the count rate) falls by half every half-life.

3 Complete ✎ the table to show the pattern in decay for isotope X in question 1.

Time in hours	0		5		10		
Activity in Bq	1800	→		→		→	
Fraction remaining	$\frac{1}{1}$	1st half-life	$\frac{1}{2}$	2nd half-life		3rd half-life	

Physics

Get back on track

Sample response

When describing the structure of the nucleus and nuclear decay, remember that:

- The nuclear model of the atom was developed because of the results of an alpha particle scattering experiment which could not be explained by earlier models.
- The changes that happen in radioactive decay should be shown in carefully balanced nuclear decay equations.
- The half-life of an isotope is the time taken for half of the atoms to decay.
- Half-life is often determined using a graph of activity over time.

Here are some exam-style questions. Use the student answers to these questions to improve your understanding of radioactive decay.

Exam-style question

1 A radioactive isotope has an initial activity of 16 000 Bq and a half-life of 30 minutes. What will the activity of the sample be after 2 hours? **(2 marks)**

0 min → 30 min → 60 → 90 → 120 is 5 half-lives. $16000 \times \frac{1}{2} \times \frac{1}{2} \times \frac{1}{2} \times \frac{1}{2} \times \frac{1}{2} = 500$ *Bq.*

The unit 'Bq' stands for becquerel. This is the number of decays each second.

Using arrows is a way to work out how many half-lives have passed.

1 What does the → represent in the student's answer? ....

2 The student's answer for the half-life is incorrect. The student counted the numbers (5) and so thought that five half-lives had passed.

a How many half-lives have really passed? Explain how you can tell. ....

b Rewrite the answer but put in activities instead of times to reach the correct answer.

....

Exam-style question

2 The isotope strontium-90 decays through β^- (beta minus) particle emission. Complete the decay equation for the decay of strontium-90. **(3 marks)**

$^{90}_{38}\text{Sr} \rightarrow {}^{90}_{37}\text{Y} + {}^{0}_{1}\text{e}$

3 **a** Circle Ⓐ the parts of the equation in the student's answer that are correct.

b Write the correct decay equation.

....

Be extra careful with β^- decay. What happens to the atomic number?

Get back on track

Your turn!

It is now time to use what you have learned to answer the exam-style question from page 129. Remember to read the question thoroughly, looking for information that may help you. Make good use of your knowledge from other areas of physics.

Exam-style question

1 A research scientist measured the count rate produced by a radiation detector for two different radioactive isotopes of the same element over a period of time.

The results are shown in Figure 1.

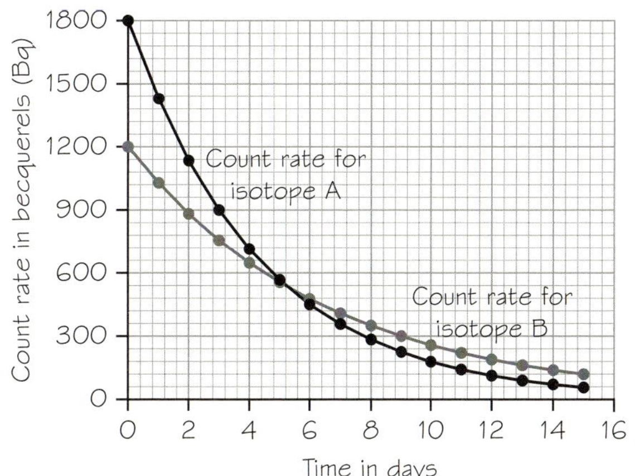

Figure 1

(a) Explain how an element can have different isotopes. **(2 marks)**

....

....

....

....

There are 2 marks here so make sure you mention one similarity and one difference between the isotopes.

(b) State which isotope was most active at the start of the experiment. **(1 mark)**

....

Use the graph. Will a more active source produce a high or low count rate?

(c) State which isotope had the longest half-life. **(1 mark)**

....

You will need to look to see which isotope's count rate halves first. Be careful as the isotopes don't start with the same level of activity.

(d) Use the graph to determine the half-life of isotope B. **(1 mark)**

....

A sample of a different radioactive isotope (iodine-53) is known to decay by β^- (beta minus) particle emission.

(e) Complete the decay equation to show the three missing values.

$$^{131}_{53}\text{I} \rightarrow {}^{\dots}_{\dots}\text{Xe} + {}^{0}_{\dots}\text{e}$$ **(3 marks)**

Watch out for the atomic number again here.

Physics

Get back on track

Need more practice?

Exam questions may ask about different parts of a topic, or parts of more than one topic. Questions about atomic structure and radioactive decay could occur as:

- questions about that topic only
- part of a question on radioactive decay and safety
- part of a question about an experiment or investigation.

Have a go at this exam-style question. 

Exam-style question

1 A research team used a mixture of isotopes as the source of alpha particles. One of these sources was the isotope radium-226.

(a) Complete the nuclear decay equation to show the alpha decay of radium-226. **(3 marks)**

$$^{226}_{88}\text{Ra} \rightarrow ^{\cdots}_{\cdots}\text{Rn} + ^{\cdots}_{\cdots}\text{He}$$

Look back at the rules for alpha decay in Skills boost 2 if you need to.

(b) The alpha particle source used in the experiment also contained small amounts of the isotope bismuth-214. This has a half-life of 20 minutes.

(i) Define the term half-life. **(1 mark)**

....

....

(ii) Determine the fraction of the original sample of bismuth-214 that will remain after 1 hour. **(1 mark)**

....

Use the → method to step through the half-lives $1 \rightarrow \frac{1}{2} \rightarrow$ etc.

Boost your grade

To boost your grade, make sure you can describe the three main types of radioactive decay and explain the changes to the nucleus using equations. Make sure you can also explain why scientists changed their model of the nucleus due to the results of the alpha particle scattering experiment and the changes to the nuclear model since then.

How confident do you feel about each of these **skills**? Colour in the bars.

1 **How do I describe the changes that happen during nuclear decay?**

2 **How do I write a decay equation?**

3 **How can I use a half-life graph to analyse radioactive decay?**

Get started

AO2/AO3

Electromagnetism

This unit will help you to understand how electricity and magnetism interact.

In the exam you will be asked to answer questions such as the one below.

Exam-style question

1 A student investigated what happens to a wire in a magnetic field. Figure 1 shows the apparatus the student used. The wire XY is connected to a switch and battery. When the switch is closed, a force acts on the wire.

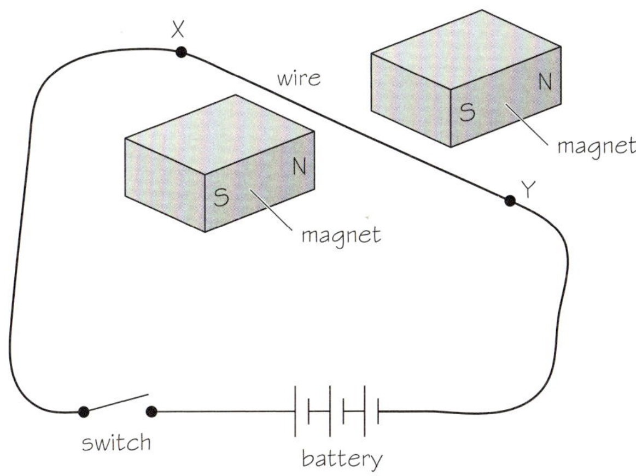

Figure 1

(a) Explain why a force acts on the wire XY when the switch is closed. **(3 marks)**

....

(b) Draw an arrow to show the direction of the force acting on the wire XY. **(1 mark)**

(c) Use an equation from the equation sheet that links the force acting on a wire carrying an electric current with magnetic flux density and current size. **(1 mark)**

....

(d) Describe **three** ways in which the force on the wire can be **increased**. **(2 marks)**

....

You will already have done some work on electromagnetism. Before starting the **skills boosts**, rate your confidence in each area. Colour in the bars.

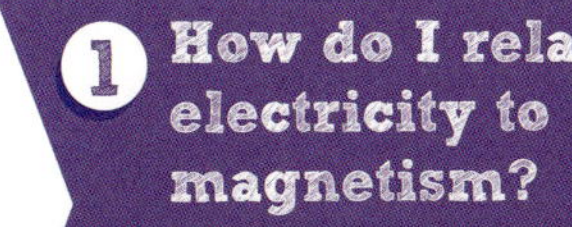

1 How do I relate electricity to magnetism?

2 What causes a force to act on a current-carrying conductor?

3 How do I calculate the force on a wire?

Physics

Get started

Magnetic field lines are a model used to represent the strength and the direction of magnetic fields. We can investigate magnetic fields using iron filings or plotting compasses. Magnetic field diagrams follow some simple rules:

- Field lines point from north (N) to south (S).
- Field lines cannot touch or cross each other.
- The closer the field lines are to each other, the higher the magnetic flux density (field strength).
- Straight, equally spaced field lines show a uniform field.

1 The diagram represents the magnetic field around a bar magnet.

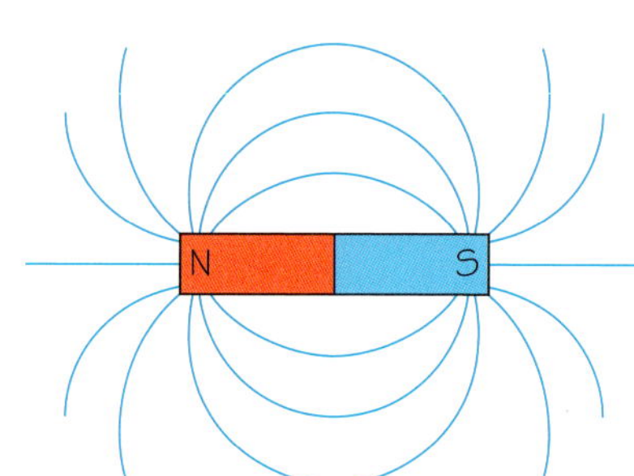

a Draw an arrow on each field line to show the direction of the magnetic field.

b Write a letter 'X' in a region of high magnetic flux density.

c Write a letter 'Y' in a region of low magnetic flux density.

If we place two magnets with unlike poles facing, we get a field diagram between the poles as shown in the diagram.

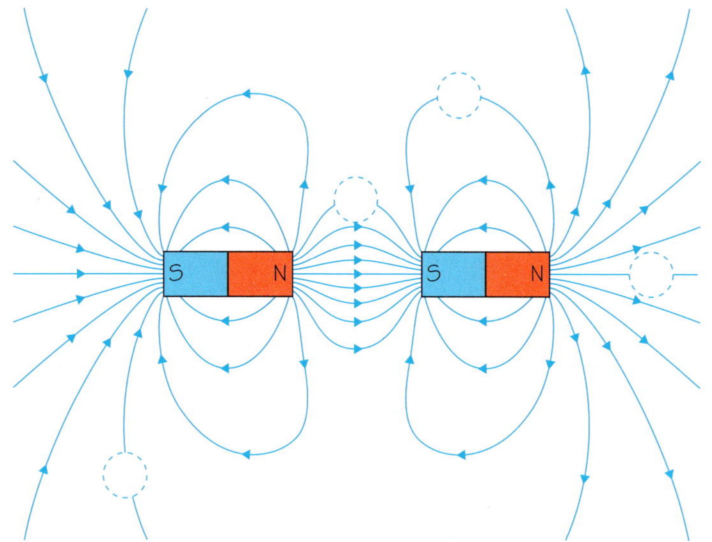

The magnetic fields of the two magnets interact so that there are some field lines between the two magnets.

2 The four circles represent plotting compasses. Add a compass needle to each to show the direction that it would point.

If we have two magnets with wide poles placed close together, the magnetic field between them looks like the diagram on the right.

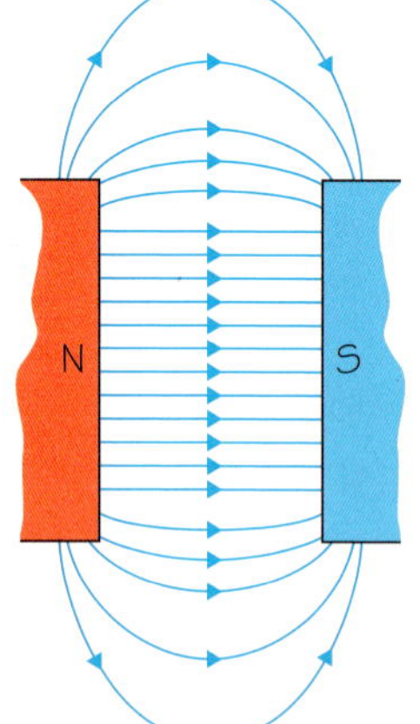

3 How would you describe the magnetic field between the magnets on the right? Tick **two** boxes.

☐ weak ☐ strong ☐ random

☐ irregular ☐ uniform

1 How do I relate electricity to magnetism?

A wire carrying an electric current produces a circular magnetic field around itself. We can predict the direction of the magnetic field using the right-hand grip rule.

1 The diagram shows the right-hand grip rule. Complete  the sentence.

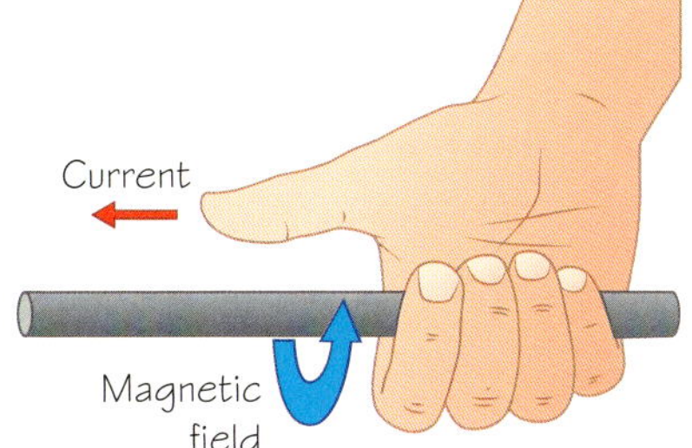

If you were to grip a wire with your right hand, your thumb would point in the direction of the .... and your fingers would point in the direction of the ....

2 The diagram shows the field around a current-carrying wire. The arrows on the field lines show the direction of the magnetic field. Draw an arrow on the diagram to show the direction of the current.

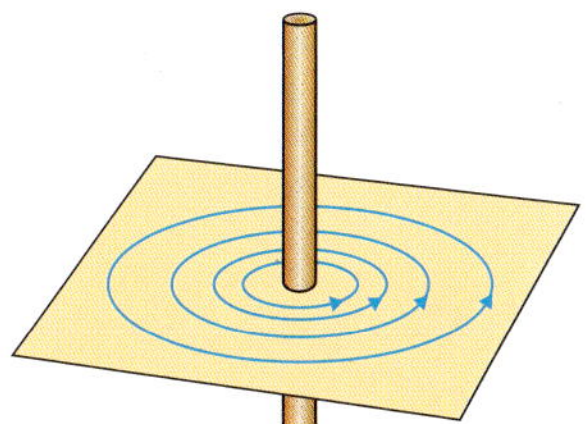

The strength of the magnetic field depends on its distance from the wire and the size of the current.

Imagine a right hand gripped around the wire.

3 Circle the correct words in **bold** to complete the sentences.

Look at the spacing of the field lines.

The magnetic field is **weakest** / **strongest** closer to the wire and gets **stronger** / **weaker** as the distance from the wire increases.

The higher the current the **weaker** / **stronger** the magnetic field.

Magnetic fields can interact with each other.

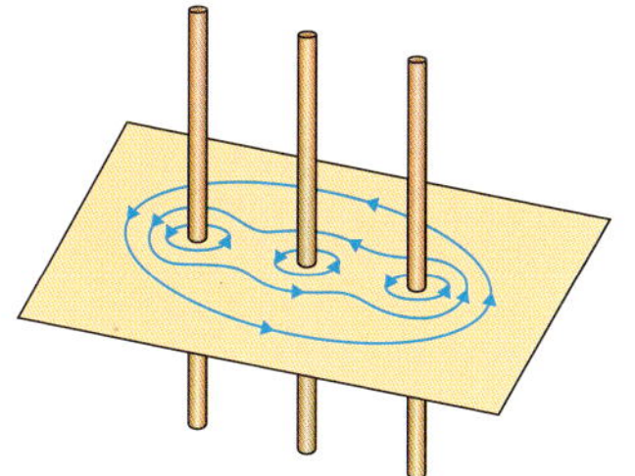

- If the magnetic fields are in the same direction, the fields add to each other and become stronger.
- If the fields are in the opposite direction, they cancel each other and become weaker.

4 The diagram shows how the magnetic fields from three parallel wires interact. Use the right-hand grip rule to predict the direction of current flow in each wire. Label each wire with an arrow to show the direction.

5 The diagram shows the magnetic fields around a solenoid and a bar magnet. Tick the correct boxes to show which statements are true and which are false.

Remember that the closer the field lines, the stronger the magnetic flux density (field strength).

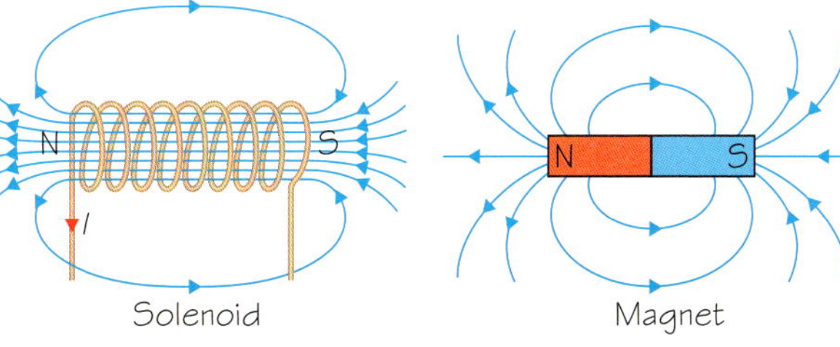

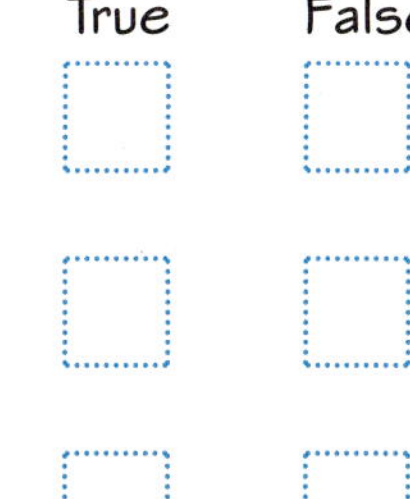

	True	False
a The magnetic fields have the same basic shape.	☐	☐
b The magnetic flux density (field strength) at the poles of the magnet is greater than the magnetic flux density at the poles of the solenoid.	☐	☐
c The magnetic flux density (field strength) at the sides of the magnet is greater than the magnetic flux density at the sides of the solenoid.	☐	☐

2 What causes a force to act on a current-carrying conductor?

The magnetic field around a wire carrying an electric current can interact with other magnetic fields. For example, it can interact with the magnetic field produced by a permanent magnet. These interactions produce forces that may cause the wire or magnet to move.

1 The diagram shows the magnetic field produced when a current-carrying wire is put between two magnets.

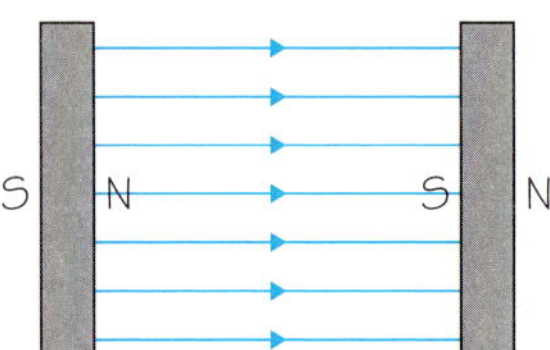

Two flat magnets produce a uniform magnetic field between them.

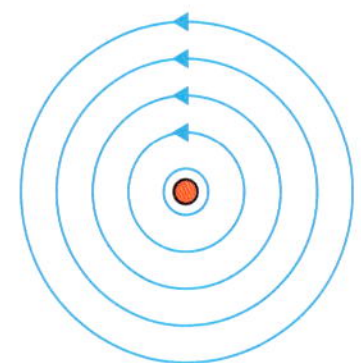
A magnetic field goes around a wire carrying a current.

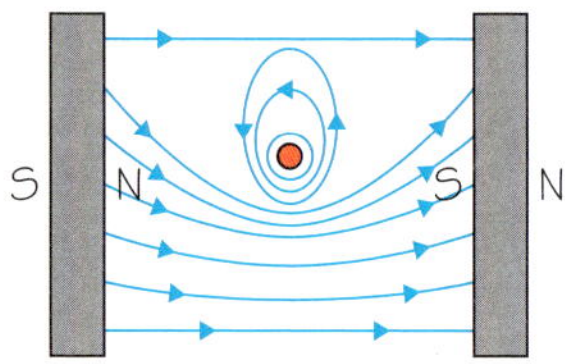

When the wire carrying a current is put between the magnets, the two fields interact to produce a force.

a Write a letter 'X' on the right-hand diagram where the magnetic field is strongest.

b Write a letter 'Y' on the right-hand diagram where the field is weakest.

c Draw an arrow on the diagram to show the direction of the force on the wire.

To predict the direction of the force on the wire, visualise the field lines as elastic bands. Imagine the elastic bands are trying to straighten by catapulting the wire away.

The wire also exerts a force on the magnets (Newton's third law).

d Circle (A) the correct words in **bold** to complete the sentences.

The force of the wire acting on the magnets is in the **same** / **opposite** direction to the force of the magnets acting on the wire.

The size of the force of the wire acting on the magnets is **the same as** / **different to** the size of the force of the magnets acting on the wire.

To create the greatest force, the magnetic field from the magnets must be at right angles to the current in the wire. Fleming's left-hand rule (shown in the diagram) gives us an easy way to predict the force (and so direction of movement) of a wire carrying an electric current.

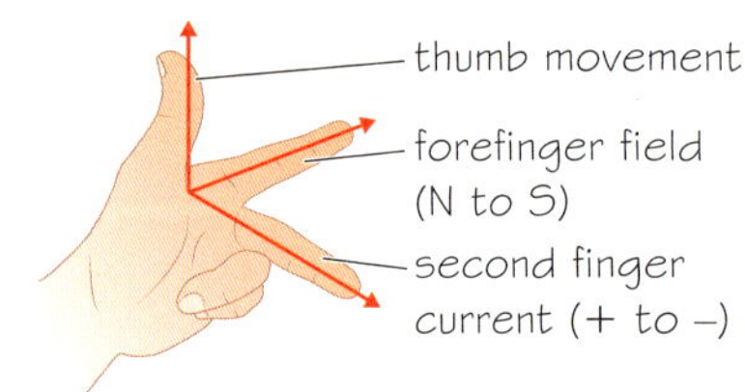

2 The diagram shows three wires in three different magnetic fields. The arrows show the direction of the current flowing in each wire.

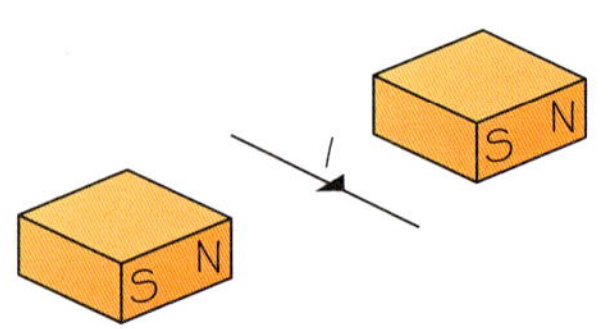

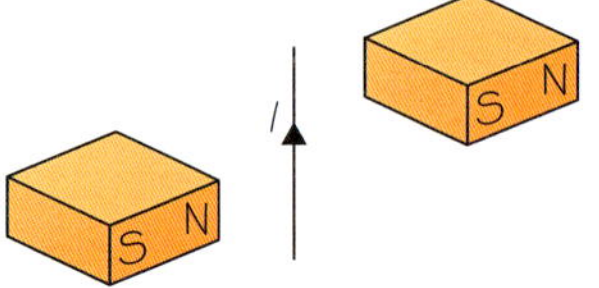

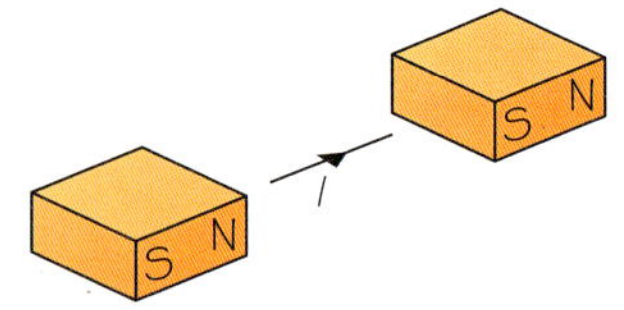

Draw arrows to show the direction of the force acting on each wire. If no force is acting, label the wire with the words 'no force'.

3 How do I calculate the force on a wire?

The force acting on a wire in a magnetic field depends on the strength of the magnetic field, the size of the current and the length of the wire carrying the current. Remember that the force is greatest when the magnetic field and wire are at right angles to each other.

We can calculate the size of the force acting on a wire using the following equation:

force on a conductor at right angles to a magnetic field carrying a current (N) = magnetic flux density (tesla, T) × current (A) × length (m)

$$F = B \times I \times l$$

You don't need to learn this equation, but you do need to use it correctly. You need to learn the physical quantities, their symbols and their units.

1 Draw lines to link each physical quantity to its symbol and unit. One has been done for you.

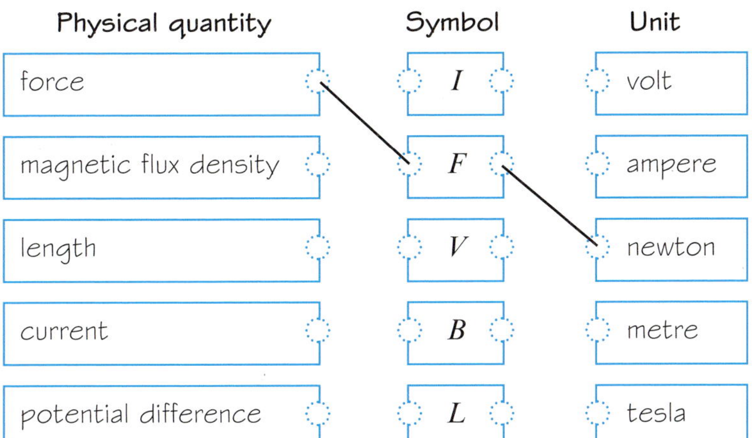

Physical quantity
Symbol
Unit
force
magnetic flux density
length
current
potential difference
I
F
V
B
L
volt
ampere
newton
metre
tesla

Remember, not all physical quantities have symbols that match the first letter of the physical quantity.

2 The diagram shows two magnets producing a strong uniform magnetic field. The north pole of each magnet is shown in red. The yoke is a metal frame which holds the two magnets.

yoke

N

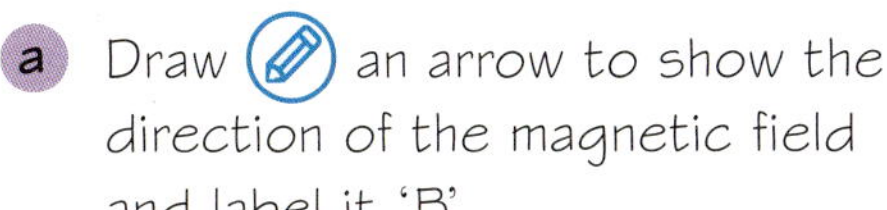

a Draw an arrow to show the direction of the magnetic field and label it 'B'.

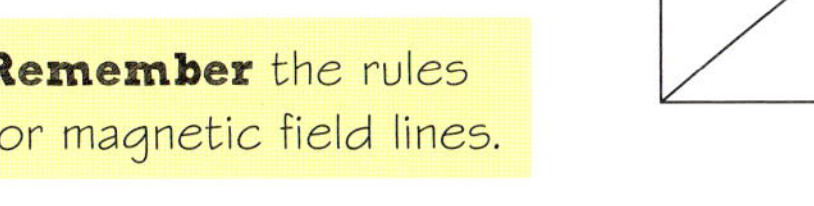

Remember the rules for magnetic field lines.

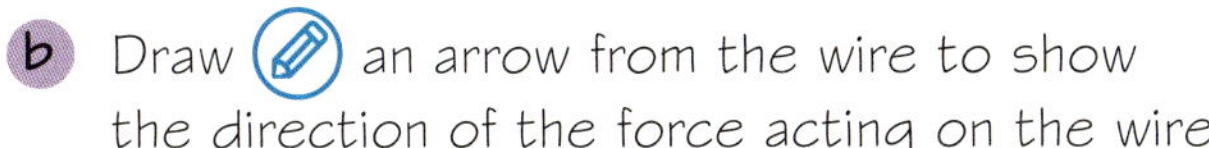

b Draw an arrow from the wire to show the direction of the force acting on the wire.

Remember to use Fleming's left-hand rule to predict the direction of the force acting on the wire.

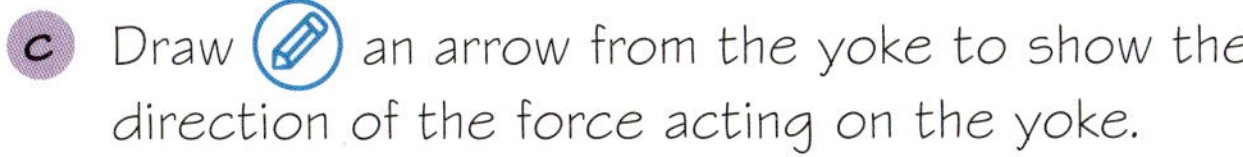

c Draw an arrow from the yoke to show the direction of the force acting on the yoke.

Remember to apply Newton's third law here.

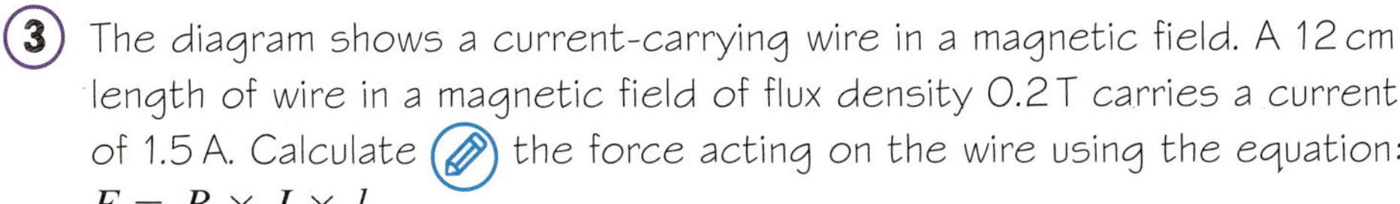

3 The diagram shows a current-carrying wire in a magnetic field. A 12 cm length of wire in a magnetic field of flux density 0.2 T carries a current of 1.5 A. Calculate the force acting on the wire using the equation: $F = B \times I \times l$

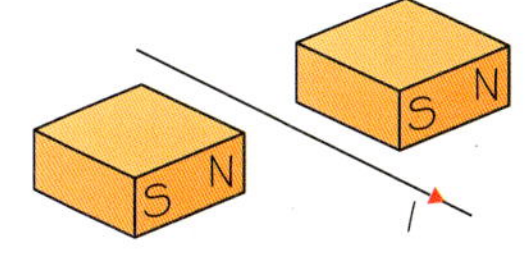

....

....

....

Remember to convert the length of the wire from centimetres to metres.

Physics

Sample response

Look at these exam-style questions and answers given by a student.

Exam-style question

1 The diagram on the left shows a vertical wire passing through a hole in a card. The wire is carrying an electric current in the direction shown by the arrow labelled *I*.

The diagram on the right shows the same card as viewed from above. Use your knowledge of electromagnetism to draw the magnetic field around the wire. **(3 marks)**

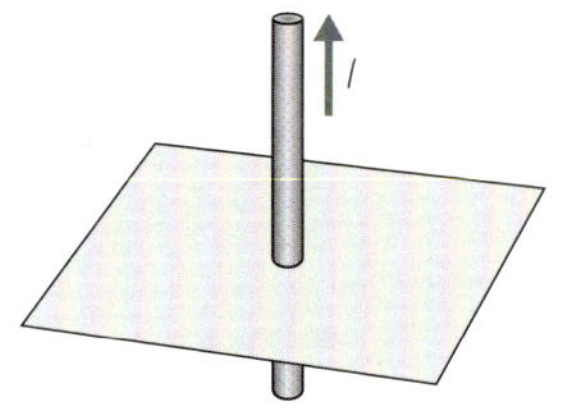

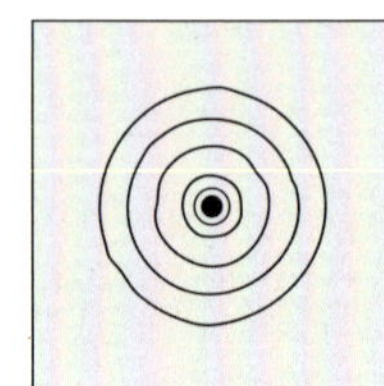

1 **a** The student's answer is shown in the diagram above. The student only gained 1 mark. How could the student have gained full marks? Draw an improved answer on the diagram on the right.

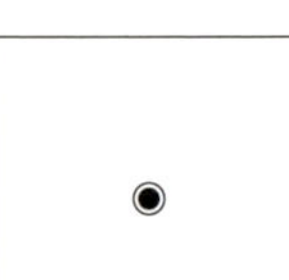

b Explain why your diagram is better than the student's diagram.

....

....

....

Exam-style question

2 The diagram shows the magnetic field around a solenoid.

(a) Write a letter 'X' in a region of high magnetic flux density. **(1 mark)**

(b) Write a letter 'Y' in a region of low magnetic flux density. **(1 mark)**

(c) Describe **two** features of the magnetic field **inside** the solenoid. **(2 marks)**

X N S Y I

Solenoid

The field lines are close together so the magnetic field is strong.

2 **a** The student has put a letter 'X' at the north pole of the solenoid to show a region of high magnetic flux density but the field lines are diverging (spreading out). Where would be a better place for the student to put the letter 'X' and why?

....

....

b There are other regions of low magnetic flux density where the student could have placed a letter 'Y'. Add a letter 'Y' to **three** more places of low magnetic flux density.

c For their answer to part **(c)**, the student only gained 1 mark. Write an improved answer which includes two features.

....

....

Your turn!

It is now time to use what you have learned to answer the exam-style question from page 137. Remember to read the question thoroughly, looking for information that may help you. Make good use of your knowledge from other areas of physics.

Remember to use Fleming's left-hand rule when answering this question.

Exam-style question

1 A student investigated what happens to a wire in a magnetic field. Figure 1 shows the apparatus the student used. The wire XY is connected to a switch and battery. When the switch is closed, a force acts on the wire.

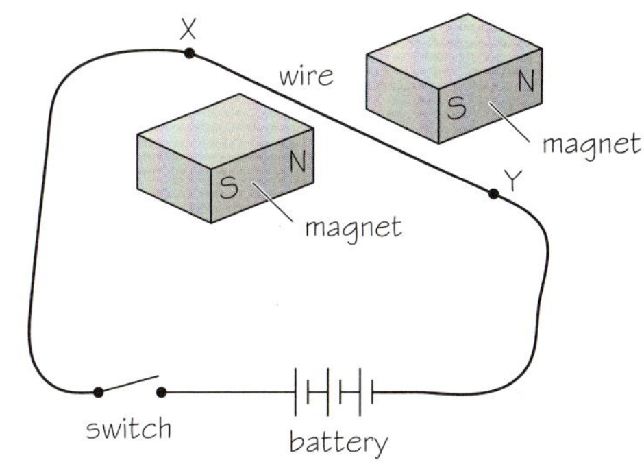

Figure 1

(a) Explain why a force acts on the wire XY when the switch is closed. **(3 marks)**

There are 3 marks available for part **(a)**. Think about three important points to write about – use the figure to help you.

(b) Draw an arrow to show the direction of the force acting on the wire XY. **(1 mark)**

Work out the direction of the force acting on the wire using Fleming's left-hand rule.

(c) Use an equation from the equation sheet that links the force acting on a wire carrying an electric current with magnetic flux density and current size. **(1 mark)**

(d) Describe **three** ways in which the force on the wire can be **increased**. **(2 marks)**

Use the equation to help you. Think about which factors on the right side of the equation could increase the force acting on the wire.

Physics

Need more practice?

Exam questions may ask about different parts of a topic, or parts of more than one topic.

Questions about electromagnetism could occur as:

- questions about that topic only
- part of a question about an experiment or investigation about devices that rely on an electromagnet to work.

Have a go at this exam-style question.

Exam-style question

1 A student wanted to measure the magnetic flux density produced by two magnets. She placed the two magnets on a top-pan balance. The top-pan balance reading changed by 16.0 g when she passed a current of 3.2 A through the wire and placed it in the magnetic field.

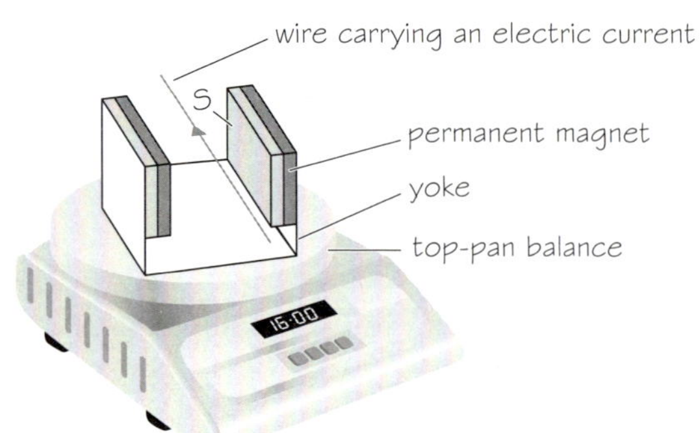

(a) Calculate the force acting on the balance. Use gravitational field strength (g) = 10 N/kg. **(2 marks)**

....

....

Remember that weight (newton, N) = mass (kilogram, kg) × gravitational field strength (newton per kilogram, N/kg), $W = m \times g$.

(b) Calculate the magnetic flux density produced by the two magnets, assuming that 0.10 m of wire is in the magnetic field. Use an equation from the equation sheet. **(4 marks)**

....

....

Boost your grade

Make sure you can use select and apply the equation: force on a conductor at right angles to a magnetic field carrying a current (newton, N) = magnetic flux density (tesla, T) × current (ampere, A) × length (metre, m), $F = B \times I \times l$.

Make sure you learn and can apply the right-hand grip rule and Fleming's left-hand rule.

How confident do you feel about each of these **skills**? Colour in the bars.

1 How do I relate electricity to magnetism?

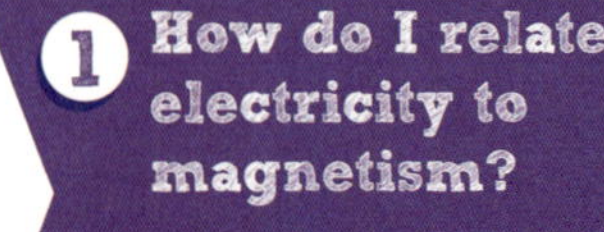

2 What causes a force to act on a current-carrying conductor?

3 How do I calculate the force on a wire?

Dealing with equations, calculations and SI units

This unit will help you to answer questions involving physics equations and calculations.

In the exam, you will be asked to answer questions such as the one below.

Exam-style question

1 A student wants to find the specific latent heat of ice by doing an experiment. The diagram shows the circuit she uses, and how she sets up the experiment.

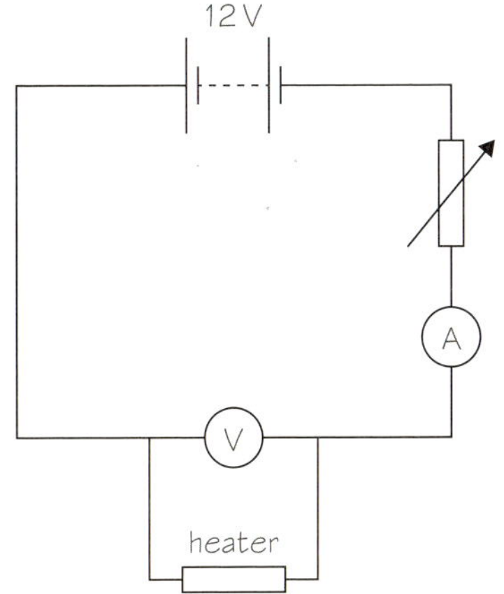

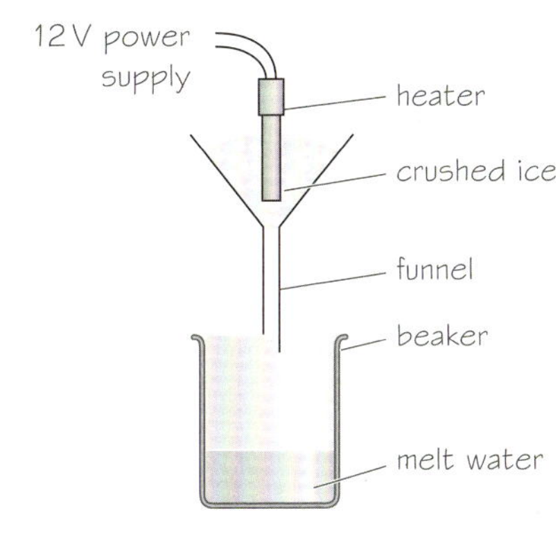

The student sets the current to 4.0 A and the potential difference across the heater to 8.0 V using the variable resistor. She runs the experiment for exactly 5 minutes.

(a) Calculate the thermal energy supplied by the heater. Use an equation from the formula sheet.

.... **(3 marks)**

(b) During the experiment, 34.3 g of melt water dripped into the beaker. Calculate the specific latent heat of fusion of water. Give your answer in J/kg.

.... **(3 marks)**

You will already have done some work on equations, calculations and SI units. Before starting the **skills boosts**, rate your confidence in equations, calculations and SI units. Colour in  the bars.

1 **How do I choose and use the correct equation?**

2 **How do I know which units to use for quantities?**

3 **How do I set out my calculations to gain full marks?**

Physics

Get started

There are many equations used in physics to describe the relationships between physical quantities. These equations may be written in words or symbols. In the exam, you are given some equations on a formula sheet, but you need to remember others.

Symbols are a way of representing physical quantities and units. For example, the equation momentum = mass × velocity is represented by $p = m \times v$.

1 In $p = m \times v$, p and v are lowercase letters. What physical quantities or units would they represent if they were uppercase?

P represents .... and ....

V represents ...., ... and ....

The International System of Units is used by all scientists. Units in this system are called SI units.

2 All the physical quantities below have an SI unit. Circle the correct SI unit for each quantity.

Distance/length **centimetre / metre / kilometre**	Time **second / minute / hour**	Energy **joule / kilojoule / megajoule**
Power **watt / kilowatt / megawatt**	Mass **gram / kilogram / tonne**	Current **milliampere / ampere**
Force **newton / kilonewton**	Potential difference **millivolt / volt / kilovolt**	Pressure **pascal / kilopascal**

Make sure you know what the correct SI units are for each physical quantity. Some equations in physics only work when SI units are used. For example, $F = m \times a$ does not work if the mass is in grams – it must be in kilograms!

The base units you will use in physics are: metre, kilogram, second, ampere and kelvin. Other units are formed from these base units and are called derived units.

For example, density $= \frac{\text{mass}}{\text{volume}}$, so the unit of density will be the unit of mass divided by the unit of volume, which is kg/m^3.

3 Derive the unit for momentum from the equation.

momentum = mass × velocity

....

You often need to rearrange an equation. For example, to calculate mass using the equation:

momentum = mass × velocity, $p = m \times v$

We need to make m the subject of the equation. The subject of the equation $p = m \times v$ is p. This is the letter on its own on one side of the equals sign.

To do this we divide both sides by v:

$$\frac{p}{v} = m \times \frac{v}{v}$$

$$\frac{p}{v} = m \times \frac{\cancel{v}}{\cancel{v}}$$

$\frac{v}{v} = 1$

leaving us with

$$\frac{p}{v} = m \text{ or } m = \frac{p}{v}$$

4 Rearrange $F = m \times a$ to make a the subject of the equation. ....

How do I choose and use the correct equation?

There are many equations that you need to use and apply in combined science physics. There are 20 that you need to remember and eight that you are given on a formula sheet.

Exam-style question

1 A lamp has a current of 1.6 A flowing through it.

Calculate the charge that passes through the lamp in 25 s. **(3 marks)**

First you need to identify the physical quantities. A physical quantity is a physical property that you can measure such as temperature, potential difference and mass.

Remember to learn the unit for each physical quantity. The unit for charge is the coulomb.

Remember A question might not state the name of the physical quantity, but it will give you the value of the quantity instead.

Remember These are the symbols for the physical quantities rather than the symbols for the units.

1 a Underline the physical quantities given in the question above.

b Circle the physical quantity you are being asked to calculate.

c Write the symbols for the three physical quantities.

....

We now need to choose an equation that contains these three quantities.

d Here are some equations related to electricity. Circle the equation that contains the three quantities.

$E = Q \times V$	$P = \frac{E}{t}$	$E = I \times V \times t$	$Q = I \times t$	$P = I \times V$
$F = B \times I \times l$	$V = I \times R$	$P = I^2 \times R$	$V_P \times I_P = V_S \times I_S$	

e Use the equation you have chosen to calculate the answer. Don't forget to give the unit!

....

....

Exam-style question

2 A resistor dissipates 2.4 W of power when a potential difference of 12 V is placed across it. Calculate the current that flows through the resistor.

The equation that links P, V and I is $P = I \times V$ but it needs to be rearranged.

2 a Rearrange $P = I \times V$ to make I the subject.

....

b Calculate the answer and give the correct unit.

....

Some equations can be more challenging than others. For example, the equation linking electrical power with current and resistance has a (current)^2 term.

As in maths, you use the order of operations (BIDMAS) to work out the right order to do your calculations: B – brackets, I – indices, D – division, M – multiplication, A – addition, S – subtraction.

2 How do I know which units to use for quantities?

Some physical quantities do not have a named unit of their own. For example, the unit for density is derived from the units of mass and volume from which density is calculated.

1 An aluminium cube of side length 0.1 m has a mass of 2.7 kg.

a Calculate the volume of the aluminium cube.

....

b Calculate the density of aluminium. Give the unit.

$$\text{density} = \frac{\text{mass}}{\text{volume}}$$

....

....

> As the SI unit of mass is kg and the unit of volume is m^3, the SI unit of density is kg/m^3.

Units are sometimes given prefixes such as mega, kilo, centi, milli, micro and nano. You need to know how to convert one unit to another. There are 1000 m in 1 km. So, to convert km to m, we need to multiply by 1000. To convert m to km, we need to divide by 1000.

2 Match the prefix to the correct multipliers. One has been done for you.

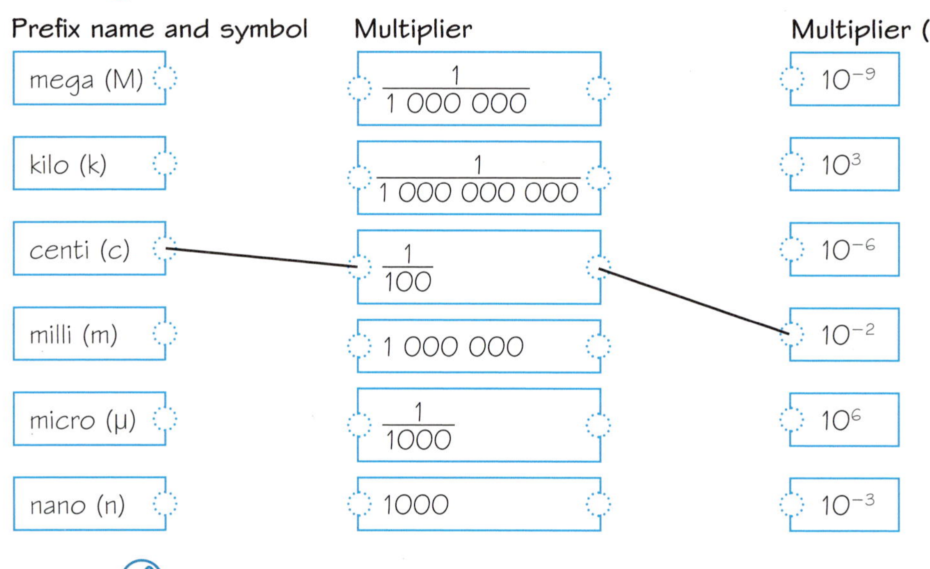

Prefix name and symbol	Multiplier	Multiplier (standard form)
mega (M)	$\frac{1}{1\,000\,000}$	10^{-9}
kilo (k)	$\frac{1}{1\,000\,000\,000}$	10^{3}
centi (c)	$\frac{1}{100}$	10^{-6}
milli (m)	1 000 000	10^{-2}
micro (μ)	$\frac{1}{1000}$	10^{6}
nano (n)	1000	10^{-3}

> Everyday use of prefixes can help you to remember whether they are big or small quantities. For example, 'mega' is used to mean 'very big' and 'micro' is used to mean very small.

3 Convert these quantities.

145 m = ... km	145 m = ... cm	2440 mm = ... m
97.7 MHz = ... Hz	48 mV = ... V	101 300 Pa = ... kPa
2300 W = ... kW		

Questions requiring calculations of speed may sometimes give units of distance in centimetres, metres or kilometres, and units of time in seconds, minutes, hours or even days or years. For example, the average speed of a glacier might be given in cm/day or m/year.

4 Calculate the missing quantities in the table.

Object	Distance	Time	Average speed
car	240 m	40 s	
lizard		15 s	20 cm/s
rocket	480 km		7.5 km/s

> For most questions using average speed, you do not need to use SI units. For questions that use velocity, such as the equations of motion, momentum and kinetic energy, you would need to convert units for distance, time and velocity to SI units.

3 How do I set out my calculations to gain full marks?

It is good practice to structure calculations in a logical way and show your working. Five logical steps are: identify information, choose equation, rearrange, put numbers in, calculate and add unit.

It is important to do this in an exam because you are more likely to get the answer right. In your exam, you usually gain full marks for a correct answer but zero marks for an incorrect answer without working. You will gain some marks for correct working even if the answer is wrong.

A mains electric hoist uses an electric motor to lift a car engine in a workshop.

Calculate the current drawn by the motor when it is working at its full power rating of 0.5 kW. Assume mains electricity is 230 V.

Look at the logical way this calculation has been laid out.

A: $P = 0.5\,kW = 500\,W$; $V = 230\,V$; $I = ?$

B: $P = I \times V$

C: $I = \frac{P}{V}$

D: $I = \frac{500}{230}$

E: $I = 2.2\,A$

Notice how the equals signs in the calculation are all aligned.

In mathematics you are taught to put the numbers in first, then rearrange the equation. However, in science you need to be able to rearrange the equations yourself.

1. Match each line in the laid out calculation above to its description below. Write the letters in the boxes. One has been done for you.

Description	Letter
Choose the right equation and write it down.	B
Calculate the answer and give the unit.	
Identify the physical quantities, making sure the units are SI units if needed.	
Put the numbers in.	
Rearrange the equation if needed.	

2. A small electrical immersion heater supplies 6.25 kJ of energy to an insulated copper block of mass 1.00 kg. The block is initially at 20 °C and the maximum temperature it reaches is 36 °C. Calculate the specific heat capacity of copper using: change in thermal energy = mass × specific heat capacity × change in temperature

3. Name and complete the five steps in the calculation in the table below.

Step	Calculation
	$\Delta Q = 6.25\,kJ = 6250\,J$; change in temperature ($\Delta\theta$) = ...; specific heat capacity (c) = ...
	$\Delta Q = m \times c \times \Delta\theta$
	$c =$ ....
	$c = \frac{6250}{1} \times 16$
	$c =$ .... unit ....

Sample response

Remember that many calculation questions are worth several marks. If you have a logical approach to answering these questions (like the five-step approach), you are more likely to gain full marks.

Look at these exam-style questions and student responses.

Exam-style question

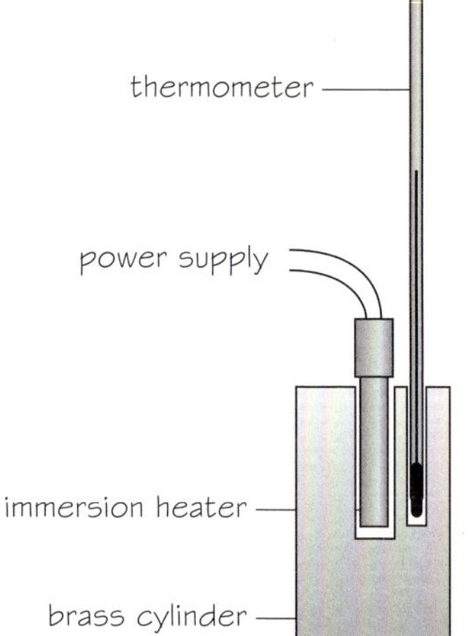

1 Harry investigates the specific heat capacity of brass.

He uses a brass cylinder with holes drilled for an immersion heater and a thermometer.

The brass cylinder has a mass of 0.50 kg.

The immersion heater has a power output of 50 W.

Harry heats the brass cylinder for 3 minutes.

(a) Calculate the energy transferred by the immersion heater in 3 minutes. 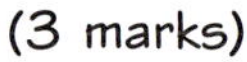 **(3 marks)**

(b) The maximum temperature rise of the brass cylinder is 45 °C.

Calculate the specific heat capacity of brass. **(3 marks)**

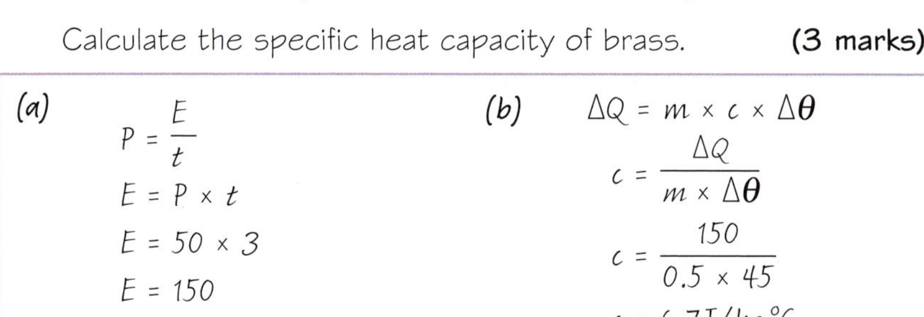

(a)
$P = \frac{E}{t}$
$E = P \times t$
$E = 50 \times 3$
$E = 150$
(b)
$\Delta Q = m \times c \times \Delta\theta$
$c = \frac{\Delta Q}{m \times \Delta\theta}$
$c = \frac{150}{0.5 \times 45}$
$c = 6.7 J/kg °C$

The student has chosen the correct equation but used it incorrectly.

1 **a** Underline values in the question with SI units.

b Circle any values in the question with non-SI units.

2 **a** The student has given the wrong answer to part (a). What two mistakes has the student made? 

....

....

b Calculate the correct value of *E*.

....

....

c The student has calculated the wrong answer for part b because they used the incorrect value for *E*.

Calculate the correct value for the specific heat capacity of brass laying out your calculation in a logical way.

....

....

....

Your turn!

It is now time to use what you have learned to answer the exam-style question from page 145. Remember to read the question thoroughly, looking for information that may help you. Make good use of your knowledge from other areas of physics.

Exam-style question

1 A student wants to find the specific latent heat of ice by doing an experiment. The diagram shows the circuit she uses, and how she sets up the experiment.

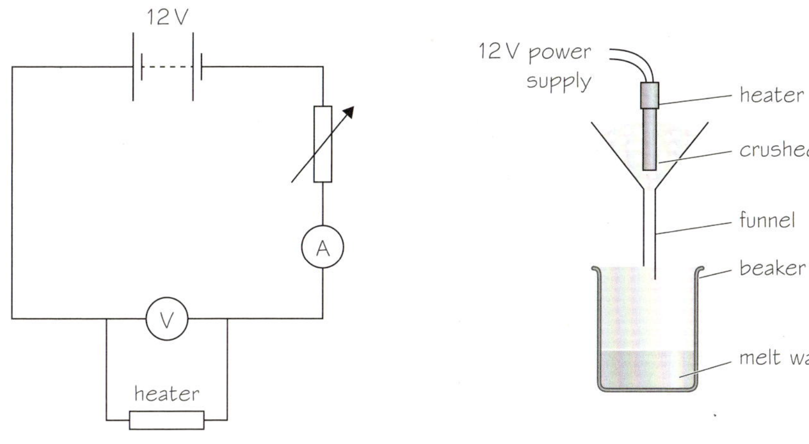

The student sets the current to 4.0 A and the potential difference across the heater to 8.0 V using the variable resistor. She runs the experiment for exactly 5 minutes.

(a) Calculate the thermal energy supplied by the heater. Use an equation from the formula sheet. **(3 marks)**

....

....

....

....

(b) During the experiment, 34.3 g of melt water dripped into the beaker. Calculate the specific latent heat of fusion of water. Give your answer in J/kg. **(3 marks)**

....

....

....

....

Remember to follow the five logical steps: identify information, choose equation, rearrange, put numbers in, calculate and add unit.

Physics

Need more practice?

Exam questions may ask about different parts of a topic, or parts of more than one topic. Questions about equations, calculations and SI units could occur as:

- part of a question in many topics
- part of a question about an experiment or investigation.

Have a go at this exam-style question.

The equation that links resistance and current with electrical power is: $P = I^2 R$

Exam-style question

1 A 50 km section of National Grid power line has a resistance of 2.7 Ω and carries a current of 5000 A.

(a) Calculate the power loss in the power line.

Give your answer in megawatts, MW. **(3 marks)**

(b) A second power line has the same resistance as the first power line and carries half the current. Describe how this affects the power loss in the second power line. **(4 marks)**

To get high marks, back up your answer to part (b) with a second calculation.

Boost your grade

To improve your grade, make sure you learn the twenty equations you need to remember. Make sure too that you learn the symbols for the physical quantities in the equations and their SI units.

Practise using the physics equations, making sure you use the five logical steps to lay out your calculations. Remember to use good maths skills such as order of operations (BIDMAS) in calculations and cross-multiplication when rearranging equations.

How confident do you feel about each of these **skills**? Colour in the bars.

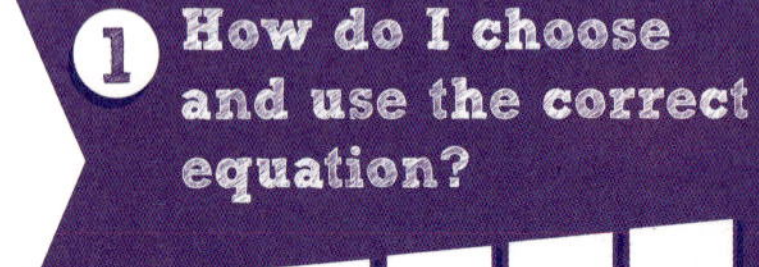

1 How do I choose and use the correct equation?

2 How do I know which units to use for quantities?

3 How do I set out my calculations to gain full marks?

6 Wave reflection, refraction and absorption

This unit will help you understand wave behaviour in terms of reflection, refraction and energy transfer.

In the exam, you will be asked to answer questions such as the one below.

Exam-style question

1 Figure 1 shows a light ray entering a semi-circular glass block from the air at point A and leaving at point B.

(a) Name the line labelled x–y on Figure 1. **(1 mark)**

....

....

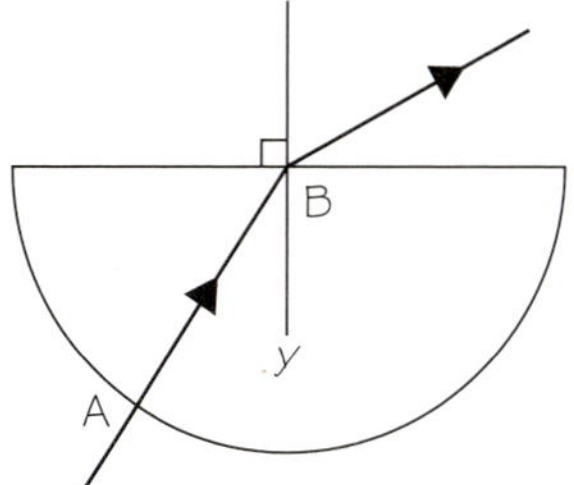

Figure 1

(b) Describe what happens to the wavelength, frequency and speed of light as the ray enters the glass at A. **(3 marks)**

....

(c) Explain why the ray of light leaving the glass block at point B carries less energy than the ray of light entering the block at point A. **(3 marks)**

....

You will already have done some work on refraction, reflection and energy transfer by waves. Before starting the **skills boosts**, rate your confidence in each area. Colour in the bars.

1 How do I explain what waves do at boundaries?

2 How do I apply ideas about absorption and emission of radiation?

3 How do I draw diagrams to explain refraction?

Physics

Get started

Waves are generated by an energy source and transfer the energy to the surroundings. Although a wave travels to transfer its energy, the particles that make up the wave only move a small amount by oscillating back and forth.

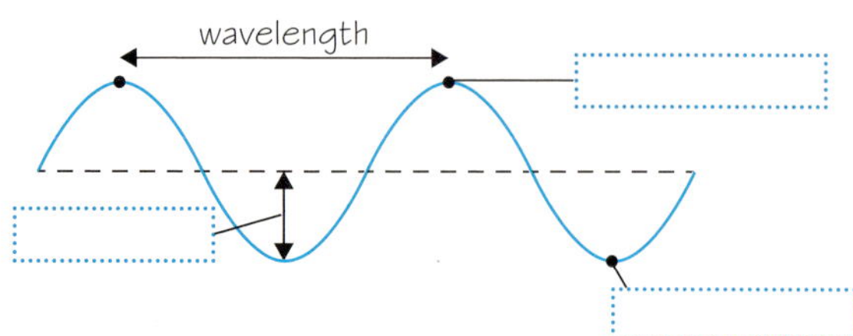

1 Label the diagram with the features of a wave shown in the word box. One has been done for you:

crest	trough	wavelength	amplitude

Wave speed describes how fast a wave travels. We can calculate wave speed from the distance a wave travels divided by the time it takes to travel that distance.

$$\text{wave speed} = \frac{\text{distance}}{\text{time}}, \; v = \frac{x}{t}$$

2 A water wave in a ripple tank travels 0.5 m in 0.8 s.

Calculate the wave speed using $v = \frac{x}{t}$

Work out the answer.

Wave frequency (f) is the number of waves passing a point each second and is measured in hertz (Hz). The relationship between frequency, wave speed and wavelength is given by the equation:

$$\text{wave speed} = \text{frequency} \times \text{wavelength}, \; v = f \times \lambda$$

3 A sound wave has a frequency of 165 Hz and a wavelength of 2.00 m. Calculate the speed of sound using $v = f \times \lambda$.

Work out the answer.

There are two types of wave: transverse waves and longitudinal waves. The diagram shows the two types. The red dot and the red line represent wave particles within the waves.

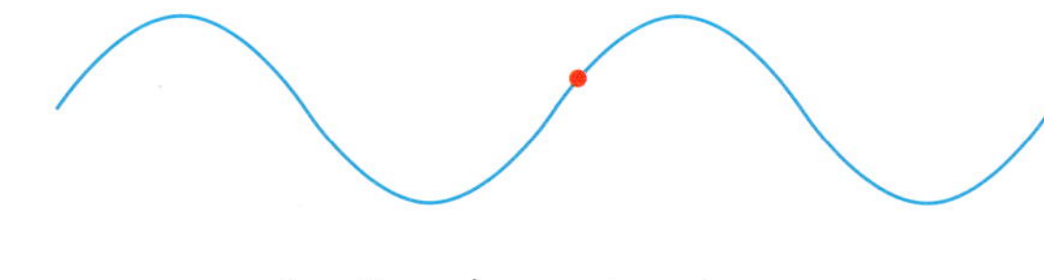

4 Tick whether each statement about the waves is true or false. The first one has been done for you.

		true	false
a	Wave A is a transverse wave.	✓	
b	The particles in wave A move up and down at right angles to the direction of wave travel.		
c	Wave B is a longitudinal wave.		
d	The particles in wave B move back and forth in the same direction of wave travel.		
e	Wave A and B have different wavelengths.		

Longitudinal means 'lengthwise'; transverse means 'across'.

How do I explain what waves do at boundaries?

Waves transfer energy away from a source. This energy can pass through a substance (transmission) or may be transferred to the substance (absorption). For example, sound waves carry energy from a loudspeaker to your ears so you can hear the sound.

When incident (incoming) waves or rays meet a boundary between one substance and another, they can bounce off the substance (reflection) or they can move from one substance to the other and may change direction (refraction).

1 Look at the diagram of a light ray hitting a glass block. Complete these sentences using words from the box.

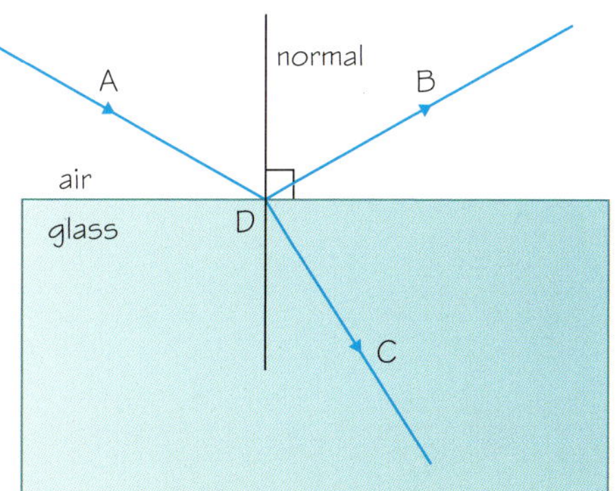

refracted	incident	absorbed	reflected

Light ray A is shone at a glass block. This is the .... ray. Some energy from ray A enters the glass and is .... to become ray C. Ray C is refracted towards the normal by the glass block because light slows down as it enters glass. Some energy from ray A does not enter the glass and is .... to become ray B. A small percentage of energy may be .... by the glass.

Remember When a wave changes speed as it moves from one substance to another, it is refracted.

A transparent medium (substance) transmits waves, but an opaque medium reflects and absorbs waves. Glass is transparent to light and infrared radiation but opaque to ultraviolet radiation.

2 Imagine you are sitting next to a glass window. Tick the **two** correct statements.

A You do get a suntan because ultraviolet light from the Sun is transmitted by the glass. ☐

B You don't get a suntan because ultraviolet light from the Sun is absorbed by the glass. ☐

C You don't get a suntan because ultraviolet light from the Sun is reflected by the glass. ☐

D You do feel warm because infrared radiation from the Sun is transmitted by the glass. ☐

E You don't feel warm because infrared radiation from the Sun is absorbed by the glass. ☐

F You don't feel warm because infrared radiation from the Sun is reflected by the glass. ☐

The further that waves travel through a substance, the greater the amount of energy that is absorbed and so less energy is transmitted.

3 At sea depths of 200 m or more, it is almost completely dark. Complete these sentences that explain why, using words from the box.

reflected	absorbed	transmitted	refracted

As light rays from the Sun reach the sea, a small percentage of light is .... but most is .... as it enters the water. As the light travels through the water, more and more light is .... and less and less is ....

2 How do I apply ideas about absorption and emission of radiation?

In an atom, electrons orbit the nucleus in shells. An electron can jump into a higher orbit (electron shell) when it absorbs electromagnetic radiation of the correct frequency. This produces a black line at that frequency in the atom's electromagnetic spectrum. Electrons emit electromagnetic radiation at the same frequencies when they jump into a lower orbit. This produces coloured lines in the electromagnetic spectrum.

1 The diagram shows the emission and absorption spectra for hydrogen in the visible region of the electromagnetic spectrum. Tick whether each statement is **true** or **false**.

Hydrogen absorption spectrum

Hydrogen emission spectrum

400 nm 656 nm 700 nm

Each wavelength of light has a unique colour.

		true	false
a	Hydrogen atoms emit and absorb visible light when electrons change orbits.		
b	Hydrogen emits visible light at four particular wavelengths.		
c	The lowest energy visible light emitted is violet light.		
d	The wavelengths of the lines in the emission spectrum exactly match the wavelengths of the lines in the absorption spectrum.		
e	The emission spectrum is a continuous spectrum.		

2 Which types of electromagnetic radiation are emitted by these devices?

Tick **one or more** boxes for each device.

Think about the properties of different types of electromagnetic radiation and how they can be detected.

	Gamma ray	X-ray	Ultraviolet	Visible light	Infrared	Microwave	Radio wave
filament lamp							
mobile phone							
neon light							
radioactive tracer							

3 Complete these sentences about what can happen to radiation from the Sun when it reaches the Earth's atmosphere. Use words from the box.

reflected refracted absorbed

When the Sun's radiation reaches the atmosphere, it slows down slightly, and so it is .... Some of the radiation is .... back into space and some radiation is .... by the atmosphere and clouds.

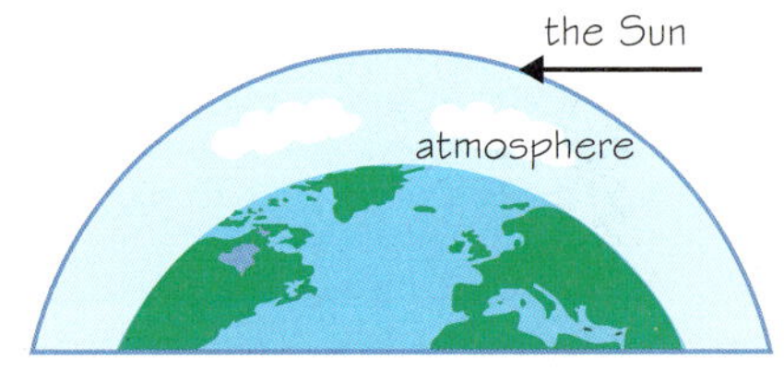

On Earth, about half the Sun's radiation that reaches the upper atmosphere reaches the surface. The rest is reflected or absorbed.

How do I draw diagrams to explain refraction?

Refraction happens as waves move from one substance into another and the wave speed changes. The frequency of the wave always remains constant. If the wave slows down, the wavelength decreases. The equation that links wave speed with frequency and wavelength is: wave speed = frequency × wavelength, $v = f \times \lambda$

We can represent a wave by a line called a wave front. You can imagine the wave front as the crest of a wave. Waves always travel at right angles to the wave front.

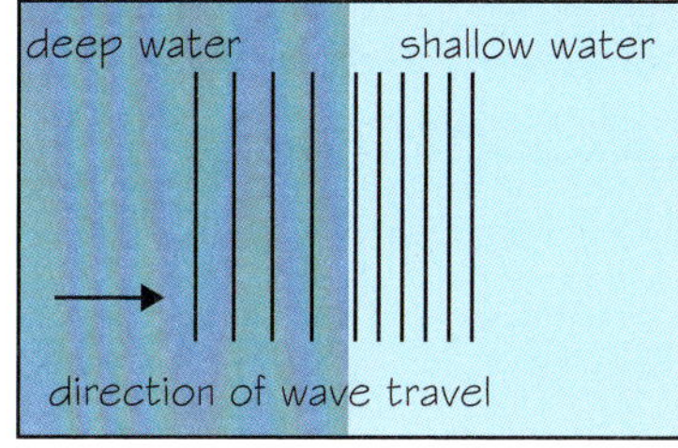

The diagram represents waves as wave fronts. It shows that water waves slow down as they reach shallower water. The frequency of the waves remains constant, but the wavelength decreases.

1 Water waves in deep water have a frequency of 2 Hz and a speed of 0.50 m/s. They reach shallow water and slow to 0.40 m/s. Calculate the wavelength of the waves in the deep water and in the shallow water. Use $\lambda = \frac{v}{f}$

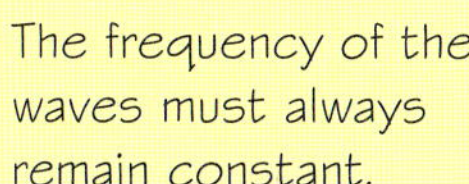

The frequency of the waves must always remain constant.

If water waves reach shallow water at an angle, they slow down and change direction. Look at the diagrams A and B.
The blue line shows the direction the water waves are travelling. The red line X–Y shows the normal. Notice how the wave direction is refracted towards the normal as the wave slows down.

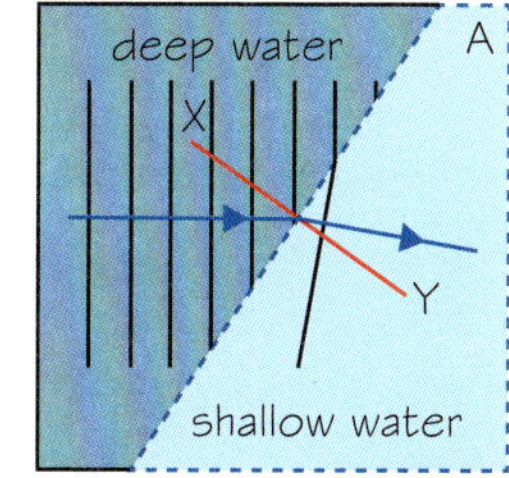

2 Some of the wave fronts in shallow water are incomplete. Draw lines to complete the wave fronts in diagram A.

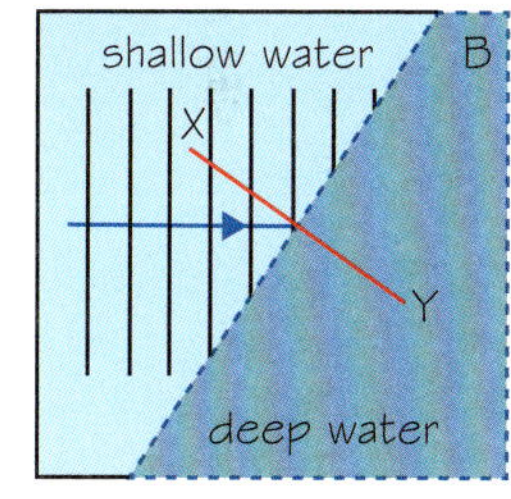

3 Diagram B shows water waves reaching deep water at an angle.

a Complete these sentences.

As waves travel from shallow water to deeper water, they .... in speed. The direction of travel is refracted away from the ....

Remember The direction a wave travels is always at right angles to the wave front.

b Draw a line on diagram B to show the direction of travel in deep water.

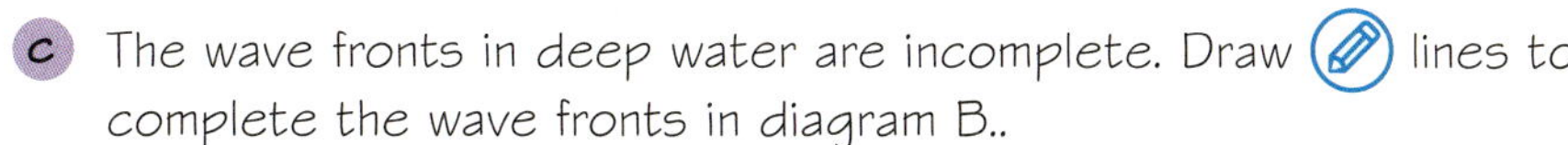

c The wave fronts in deep water are incomplete. Draw lines to complete the wave fronts in diagram B..

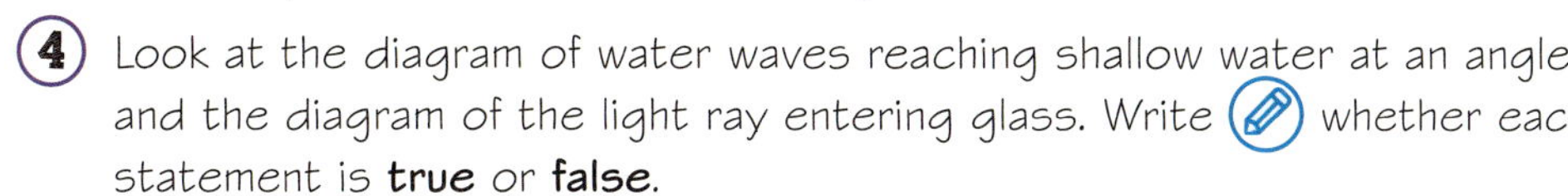

4 Look at the diagram of water waves reaching shallow water at an angle and the diagram of the light ray entering glass. Write whether each statement is **true** or **false**.

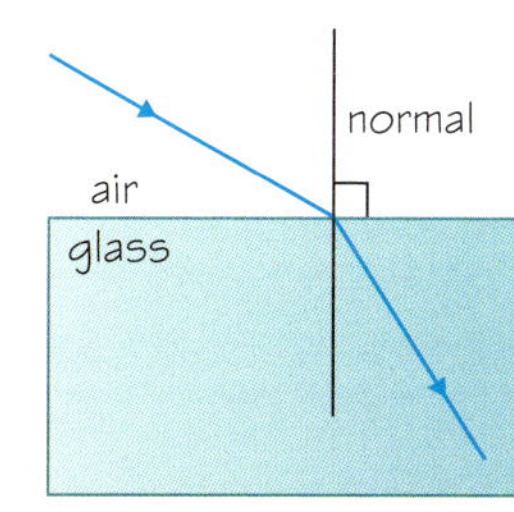

a Both change speed at the boundary. ....

b Both change frequency at the boundary. ....

c Both change wavelength at the boundary. ....

d Both change direction of travel at the boundary. ....

Physics

Sample response

Use these example student responses to improve your understanding of how to gain more marks in questions about waves.

Exam-style question

1 A student investigates water waves using a ripple tank. Figure 1 shows a wave diagram.

(a) Label the amplitude and wavelength. **(2 marks)**

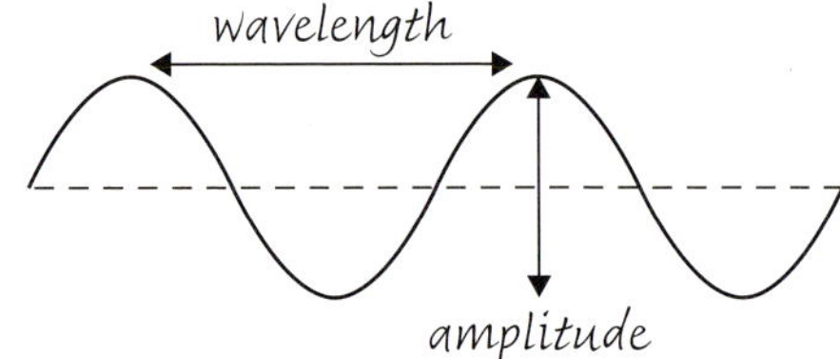

Figure 1

Figure 2 shows the student's ripple tank. The wooden bar vibrates and makes ripples. The light source projects an image of the ripples onto the screen below the ripple tank.

(b) Calculate the wavelength of the waves as seen on the screen. **(2 marks)**

There are 11 waves in 55cm. $\lambda = \frac{55}{11} = 5\text{cm}$

(c) The student measures the frequency of the waves on the screen as 16Hz. Calculate the speed of the waves. **(3 marks)**

$v = f \times \lambda$

$v = 16 \times 5$

$v = 80\,m/s$

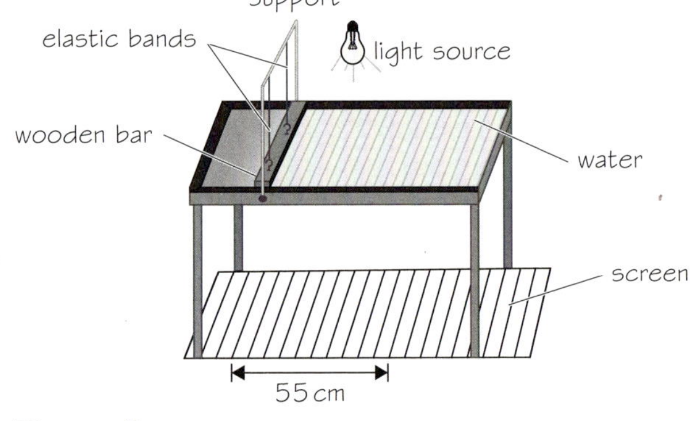

Figure 2

1 Describe the **two** mistakes the student has made in labelling the amplitude and wavelength on the wave diagram for part (a).

....

....

2 The student counted 11 waves for part (b). Write how many waves they should have counted.

....

3 The answer to part (c) is incorrect but the student gained 2 marks.

a Describe what the 2 marks were awarded for.

....

....

b Explain why a mark was not awarded for the unit.

....

....

Your turn!

It is now time to use what you have learned to answer the exam-style question from page 153. Remember to read the question thoroughly, looking for information that may help you. Make good use of your knowledge from other areas of physics.

Exam-style question

1 Figure 1 shows a light ray entering a semi-circular glass block at point A and leaving at point B.

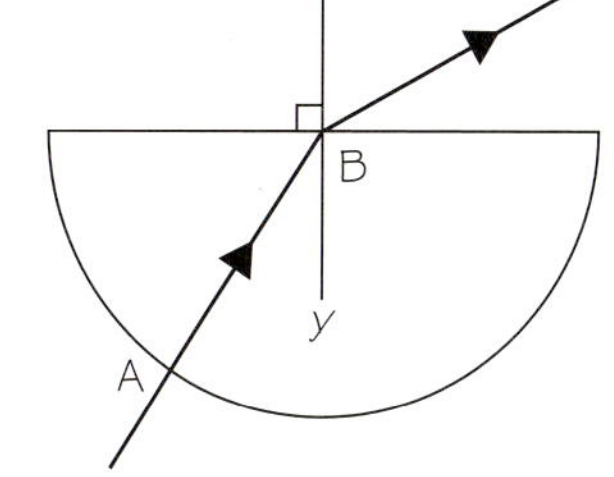

Figure 1

(a) Name the line labelled *x–y* on Figure 1. **(1 mark)**

....

....

(b) Describe what happens to the wavelength, frequency and speed of light as the ray enters the glass at A. **(3 marks)**

....

....

....

....

(c) Explain why the ray of light leaving the glass block at point B carries less energy than the ray of light entering the block at point A. **(3 marks)**

....

....

....

Physics

Need more practice?

Exam questions may ask about different parts of a topic, or parts of more than one topic.

Questions about waves could occur as:

- questions about light, electromagnetic spectrum, sound or water waves
- part of a question on how energy is transferred
- part of a question about an experiment or investigation.

Have a go at this exam-style question.

Exam-style question

1 When a light ray is shone on a glass block, not all of the energy is transferred by reflection and refraction.

(a) Describe what happens to the remaining energy. **(1 mark)**

....

....

(b) Draw and label the normal line, reflected ray and refracted ray on Figure 1. **(3 marks)**

(c) Skiers need to wear special sunglasses or goggles.

Explain why, in terms of reflection and absorption of light. **(3 marks)**

....

....

....

air

glass

Figure 1

....

Boost your grade

Make sure you practise labelling wave diagrams and learn these equations: $v = f \times \lambda$ and $v = \frac{x}{t}$
Learn the similarities and differences of longitudinal and transverse waves, including these common properties:

- waves transfer energy from a source to the surroundings
- energy from waves can be reflected, transmitted or absorbed by a substance
- waves may change direction by refraction as they move from one substance to another.

How confident do you feel about each of these **skills**? Colour in the bars.

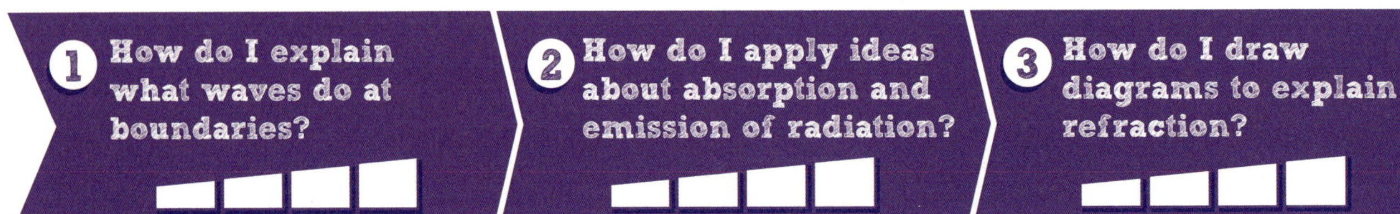

1 How do I explain what waves do at boundaries?
2 How do I apply ideas about absorption and emission of radiation?
3 How do I draw diagrams to explain refraction?

7 Answering extended response questions

This unit will help you learn more about how to answer exam questions which involve more extended writing. These questions usually award six marks and require either the detailed understanding of a process or the drawing together of ideas from several areas of physics. Extended response questions often appear as the final part of a multi-part question. The earlier parts of the question can provide information which will help in understanding the question.

In the exam, you will be asked to answer questions such as the one below.

Exam-style question

1 Figure 1 shows the structure of a part of the National Grid used to link a power station to distant homes.

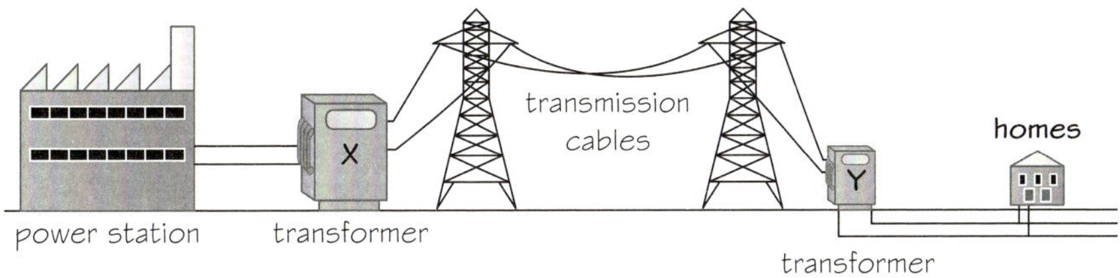

Figure 1

(a) Label the step-up and step-down transformers in Figure 1. **(1 mark)**

(b) Explain why very high potential differences are used to transfer electrical power in overhead power cables, but lower potential differences are used in houses. Your explanation should include relevant physics equations. **(6 marks)**

....

You will already have done some work about how to answer extended response questions. Before starting the **skills boosts**, rate your confidence in each area. Colour in the bars.

1 **How do I know what the question is asking me to do?**

2 **How do I organise my answer?**

3 **How do I choose the right detail to answer the question concisely?**

Extended response questions use different **command words**. These words tell you how to answer the question, for example, whether to write a one-word answer or to provide a detailed explanation.

1 The table below shows five question parts related to electrical resistance.

a Underline the command words.

b Draw lines to connect each question part to the correct meaning of the command word and then to the correct style of answer.

Question parts	Meaning	Style of answer
Explain why this circuit could be used to measure resistance …	Apply your knowledge and understanding to a new situation.	The current in the wire is causing it to heat up and so its resistance is increasing.
Evaluate the use of radioisotopes such as iodine-131 …	Give similarities and differences between several things, not just one.	Both water waves and sound waves need a medium to travel through. Water waves are transverse and sound waves are longitudinal.
Compare and contrast the properties of water waves and sound waves.	Look at the information in the question and bring it together to make a decision and come to a conclusion with evidence from the question.	This circuit can be used to measure resistance because …
The current in the wire decreases when the circuit is left on. Suggest an explanation for why this reading changes.	Say how or why something happens – 'because' will be an important part of your answer.	Radioisotopes, such as iodine-131, may emit harmful gamma radiation but …

Extended response questions require **planning** as the answer has to contain the correct scientific information in a clear and logical order.

2 A student plans an experiment to find out if different coloured surfaces affect the rate at which they emit infrared radiation and cool down. Her plan, below, contains the scientific information for a full answer, but it is not in a clear order.

Number the stages of the method so that they are in a logical order. The first is done for you.

Stage	Description
1	*Three boiling tubes, one painted black, one white and one silver are placed in a test tube rack.*
	Using a stopwatch, measure the temperature of the water in the tube each minute for ten minutes and record this data in a table.
	Repeat the process with the remaining two boiling tubes.
	Place the thermometer into the black boiling tube. Pour 20cm^3 of boiling water into the boiling tube.
	Compare the cooling of the tubes – the one which has cooled most during the ten minutes is the best emitter of infrared.
	Use the thermometer to measure the temperature of the water in the boiling tube and wait until the temperature reaches 80 °C.

When you write a method for an experiment, the information given needs to be clear enough for someone to follow the method and get the results that you expected.

 How do I know what the question is asking me to do?

Identifying the command word and making sure you know what it means are essential skills.

Exam-style question

1 Figure 1 shows how a student placed two magnets with opposite poles facing on a top-pan balance. She zeroed the balance.

She clamped a horizontal wire between the poles of the magnets, so the wire could not move.

She passed an electric current through the wire.

She observed that the top-pan balance reading changes.

Explain why a force acts on the balance and how the force can be changed. **(6 marks)**

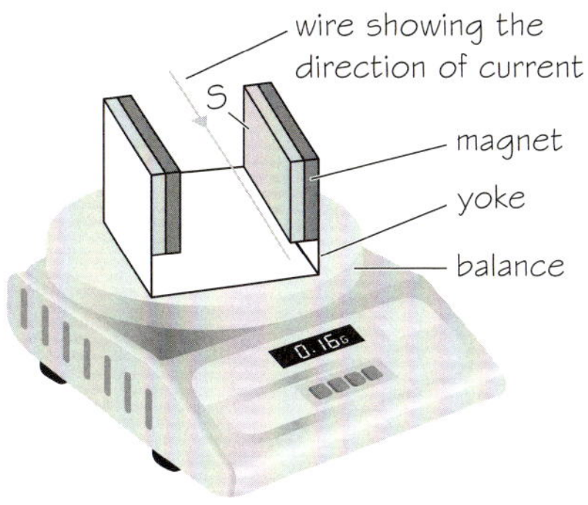

wire showing the direction of current
S
magnet
yoke
balance
0.16g

Figure 1

The balance reading changes because of an additional force.

1 **a** Circle the command word. What does the command word mean?

....

b Highlight any useful information in the question and on the diagram.

c The question is asking for two explanations, underline what they are.

In an 'explain' answer, there are usually cause and effect statements. An effect statement is usually linked to a cause statement by 'because'. Sometimes, you might use 'so' or 'as'.

2 Draw lines to link each effect to a cause. One has been done for you.

Effect

- A There is a magnetic field created around the wire …
- B There is a force acting downwards on the magnets …
- C A force acts on the wire …
- D There is a uniform magnetic field created between the magnets …
- E The force increases if the current in the wire increases …
- F The force changes direction if the current in the wire changes direction …
- G The force increases if the magnets are placed closer together …

Cause

- a because the magnetic field from the wire interacts with the magnetic field from the magnets.
- b … because there is an electric current in the wire.
- c … because the magnets have opposite poles facing.
- d … because a force acts upwards on the wire.
- e … because the strength of the magnetic field increases and it is directly proportional to the force acting on the wire.
- f … because the force exerted by the magnetic field on the wire is directly proportional to the current in the wire.
- g … because the direction of the force depends on the direction of the current and the magnetic field. It can be predicted by Fleming's left-hand rule.

Skills boost

2 How do I organise my answer?

Make sure you are clear about what the question is asking you to do from the command word. You also need to think about which physics ideas are useful for the answer.

Consider this exam-style question again.

Exam-style question

1 Figure 1 shows how a student placed two magnets with opposite poles facing on a top-pan balance. She zeroed the balance.

She clamped a horizontal wire between the poles of the magnets, so the wire could not move.

She passed an electric current through the wire.

She observed that the top-pan balance reading increased.

Explain why a force acts on the balance and how the force can be changed. **(6 marks)**

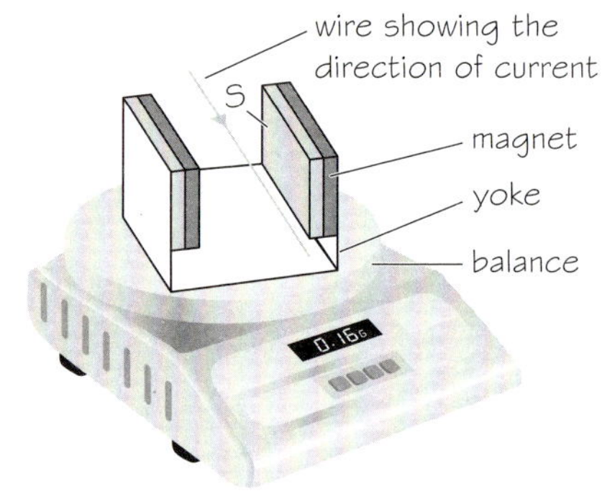

Figure 1

Before starting to write your answer, it's a good idea to think about the physics topic the question relates to.

1 Circle Ⓐ the physics topic that is covered in the question.

Energy stores and transfers	The motor effect
Resistance and Ohm's law	Newton's second law

It is important to recognise the areas of physics you will use, but you should not write everything you know about that topic.

2 Here are some notes you might use in a concise plan. Number them in a logical order.

- [ ] *Fleming's left-hand rule*
- [ ] *force increases if magnets closer together, so strength of magnetic field increases*
- [ ] $F = B \times I \times l$
- [ ] *upwards force on wire (Fleming's left-hand rule)*
- [ ] *magnetic field between magnets*
- [4] *downwards force on magnets (resultant forces)*
- [1] *current on wire – magnetic field*
- [ ] *force increases if current increases*
- [ ] *extra downward force on balance*
- [ ] *direction of current changes direction of force*
- [ ] *two magnetic fields interacting – force*

Look carefully at the question to make sure you understand what you need to do. This question has two parts, so you need to explain the force on the balance first and then how to change the force.

Make sure you include any appropriate terms and/or equations.

3 Using the cause and effect statements from skills boost 1 and your plan, write your answer to the exam-style question on a separate piece of paper.

How do I choose the right detail to answer the question concisely?

In your answer, write one or two sentences about each idea.

Exam-style question

1 A ray of light is shone at a rectangular glass block at a shallow angle as shown in Figure 1.

Explain what will happen to the energy transferred by the ray of light at the air-glass boundary, inside the glass block and at the glass-air boundary in terms of reflection, refraction, absorption and transmission of light. **(6 marks)**

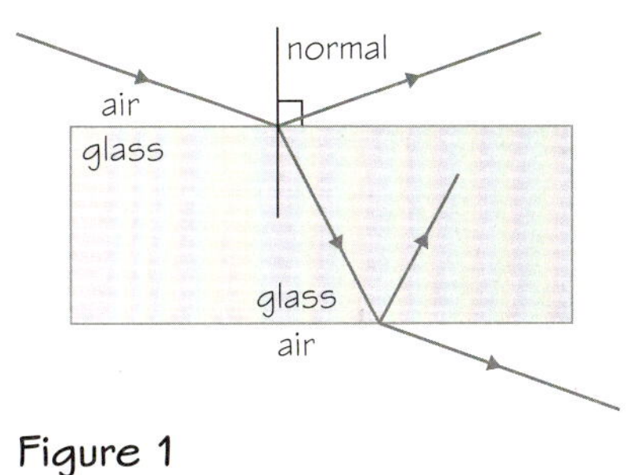

Figure 1

1 Identify the information given in the question.

a Circle the **three** important regions of the diagram that are named in the question and underline their names in the question.

b Highlight the **four** wave effects that are named in the question.

2 Identify which physics topics are needed to answer the question.

Circle the **three** main physics topics needed to answer the question above.

Thermal properties of materials	The law of conservation of energy	Reflection and refraction
Absorption and emission of wave energy	Velocity and acceleration	Particle theory

Think about the law that can explain what happens to the energy from a light ray. Can the energy be lost or destroyed?

To answer the question accurately and concisely, you need to break the question into parts and think about what is happening in each part. For this question it is the light energy in each region.

3 a Complete the table using the diagram to help you. One box has been completed for you.

Region	Reflection	Refraction	Absorption	Transmission
air-glass boundary	some energy is reflected			
inside the glass block				
glass-air boundary				

b Now write about what is happening in each region by completing the sentences below. One has been done for you. Remember to think about cause and effect using 'because'.

As the ray of light reaches the air-glass boundary, some light energy is reflected but the remaining light energy is refracted at the surface and enters the block **because** all energy from the ray must be refracted or reflected.

As the ray of light passes through the glass block ....

....

As the ray of light reaches the glass-air boundary ....

....

Sample response

To answer an extended response question, you need to:

- analyse the question to decide what the question is asking for
- identify the scientific ideas that are relevant
- put the ideas in a logical order and make connections between them
- give the right amount of detail in your answer without writing too much.

Look at this exam-style question and student response.

Exam-style question

1 A frame holding two permanent magnets is placed onto a top-pan balance as shown in Figure 1.

The top-pan balance shows the weight of the magnets and frame. A copper rod is positioned so that it is held in a fixed position between the magnets. An electric current is passed through the rod in the direction shown.

Predict whether the reading on the scale of the top-pan balance decreases or increases when there is a current in the rod. Explain your prediction. **(6 marks)**

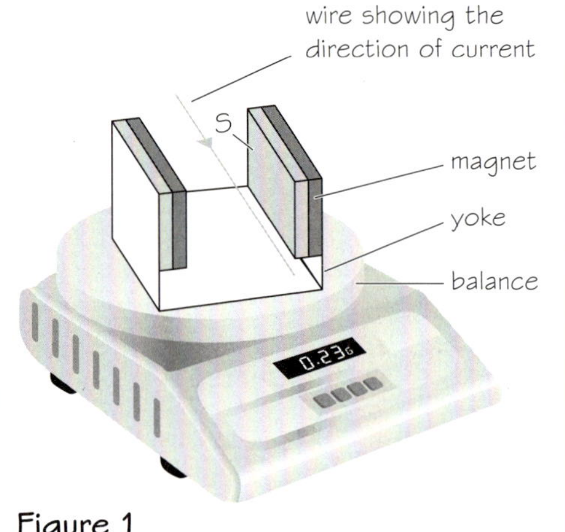

Figure 1

The 'explain' part of the question is the main part.

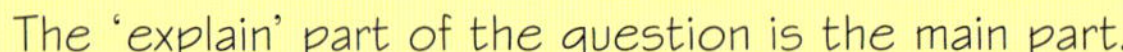

I think it will decrease. Copper is a good conductor but is not normally magnetic but the current in the rod is making it an electromagnet as it passes through. This magnetic rod is then affecting the magnets because there is a force whenever two magnets are placed near to each other because of their magnetic fields. The magnets act to repel each other so the rod is pushed away from the permanent magnets. When magnets put a force on the rod it is the same force that pushes back on the magnets and then the top-pan balance. If you make the current in the rod larger then the reading on the balance will change more because there will be an even larger force because of the equation $F = B \times I \times l$.

1 Cross out (~~cat~~) any irrelevant information the student included.

The student's answer needs to be more precise and use the correct terms. There is also information missing.

2 a Highlight the sentences where the student attempts to explain why there is a force acting on the wire.

b Underline (A) the sentence where the student attempts to describe the pair of forces acting between the magnets and the rod.

3 Use your knowledge to plan and rewrite the student answer so that it gains full marks. Use a separate sheet of paper.

Your turn!

It is now time to use what you have learned to answer the exam-style question from page 161. Remember to read the question thoroughly, looking for information that may help you. Make good use of your knowledge from other areas of physics.

Exam-style question

1 Figure 1 shows the structure of a part of the National Grid used to link a power station to distant homes.

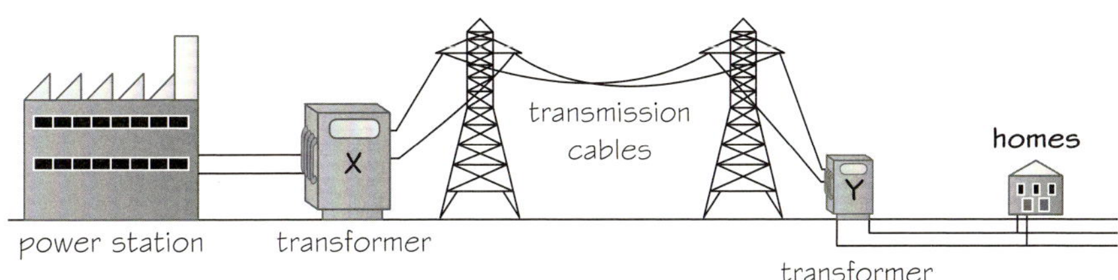

Figure 1

(a) Label the step-up and step-down transformers in Figure 1. **(1 mark)**

(b) Explain why very high potential differences are used to transfer electrical power in overhead power cables, but lower potential differences are used in houses. Your explanation should include relevant physics equations. **(6 marks)**

1 Circle the command word in part (b).

What does the command word mean? Keep that as the focus of your answer.

2 To make sure you fully understand the question, underline the information in the question you have to explain.

You need to identify any relevant equations you might need. The equation that shows electrical power transmitted by a wire or cable is: power = current × potential difference, $P = I \times V$.

3 Explain what $P = I \times V$ tells you about how to transmit a large electrical power.

....

The equation that links the current and resistance of a wire to the power it wastes due to heating is: power = current^2 × resistance, $P = I^2 \times R$.

4 Explain what this equation tells you about how to reduce the power wasted due to electrical heating.

....

5 Compare your answers to parts 3 and 4 and explain why high voltages are used to transmit power in long cables.

....

....

6 Write a plan for your answer.

....

....

7 Now write your answer on a separate piece of paper. Use the information in your answers in questions 1 to 6 to help.

Physics

Need more practice?

In an exam, extended response questions could occur as:

- simple standalone questions
- part of a question about any topic you have studied
- part of a question about an experimental procedure.

Have a go at this exam-style question. If you need more space to write your answer, continue on a separate piece of paper.

Exam-style question

1 There are many different types of wave including sound waves, water waves and electromagnetic waves.

Some wave properties are common to all waves. Other wave properties are unique to certain types of wave.

Compare and contrast the properties of electromagnetic waves and sound waves. **(6 marks)**

When you compare and contrast, you need to give similarities and differences between two or more things.

Circle the command words in the question, then underline the things you need to compare and contrast. This will help you to form a plan.

Think about the similarities of the two types of wave – these are the common properties of waves. List three or four if you can.

Think about the differences between the two types of wave. Again, list three or four if you can.

Boost your grade

To boost your grade, make sure that you describe all of the relevant points clearly using the information provided and your physics knowledge. You should annotate diagrams to help you understand them and plan your answer so that you make the required number of scientific points.

How confident do you feel about each of these **skills**? Colour in the bars.

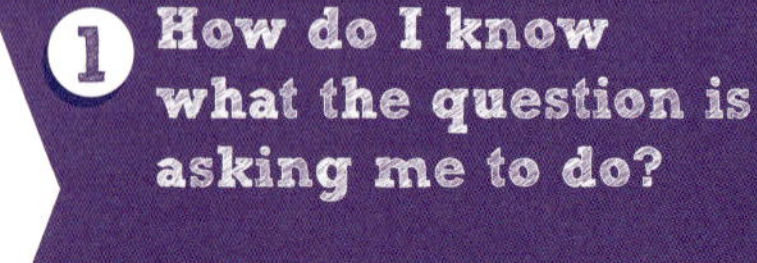

1 How do I know what the question is asking me to do?

2 How do I plan my answer?

3 How do I choose the right detail to answer the question concisely?

Answers

Biology

Unit 1

Page 2

1 a carbon dioxide – higher concentration in capillary than in alveolus

oxygen – higher concentration in alveolus than in capillary

b Carbon dioxide moves from capillary to alveolus

c Oxygen moves from alveolus to capillary

2 Osmosis is the movement of water molecules across a semi-permeable membrane. They move from a region of lower solute concentration to a region of higher solute concentration where water molecules are less concentrated.

3 a Arrow from surrounding soil into root hair cell

b Arrow from surrounding soil into root hair cell

c lower; energy; against

Page 3

1

Block	Surface area (cm^2)	Volume (cm^3)	SA : V	Order of colour change
A	6 (= 1 cm × 1 cm × 6 sides)	1 (1 cm × 1 cm × 1 cm)	6 : 1	1st
B	(2 × 2 × 6) = 24	(2 × 2 × 2) = 8	24 : 8 or 3 : 1	2nd
C	(3 × 3 × 6) = 54	(3 × 3 × 3) = 27	54 : 27 or 2 : 1	3rd

2 a A: 0.5; B: 1; C: 2

b C; concentration; greatest / steepest; fastest

Page 4

1 out of; shrink

2 solution A – lower than
solution B – equal to
solution C – higher than

3 The sprinkled sugar creates a higher solute concentration outside the strawberry cells than inside. Water molecules move from a low solute concentration to a higher solute concentration, so they move out of the strawberry cells into the bowl and mix with the sugar to create a syrup.

4 Arrows from each cell to show [the **net** direction of water molecule movement between cells and into or out of the solution, based on the solute concentrations shown]

Page 5

1 C – to the xylem cells
B – root hair cell membrane containing transport proteins
A – concentration gradient
E – low concentration of mineral ions in the soil
D – higher concentration of mineral ions in the root hair cell

2 a more; lower; against; active transport; transport proteins; energy

b A – active transport
B – transport protein
C – cells supplying energy

Page 6

1 Because they stated that excess water would affect / increase the mass of the cube.

2 By stating that this is because there is no net movement of water molecules.

3 Repeat the test using concentrations between 10 and 30 g/dm^3.

Page 7

1 Prediction: tube A will be larger / grown bigger / gained mass (1).

Explanation: it will have taken in water molecules by osmosis (1). The water molecules will have moved from an area of (higher water potential / lower solute concentration) in the 5% sucrose solution to an area of (lower water potential / higher solute concentration) in the 20% sucrose solution (1) across a partially / semi-permeable membrane (1)

Page 8

1 Waste carbon dioxide molecules move from a higher concentration inside the earthworm to the surrounding air. Oxygen molecules move from high concentration in the surrounding air through the earthworm skin to a lower concentration in the cells (1) by diffusion (1).

2 Pure water has zero solute concentration. (1) Concentration gradient is very steep from pure water to red blood cell. Water molecules rapidly move into the red blood cell until the cell membrane is ruptured / damaged / lysed (1) by osmosis (1).

3 By active transport (from an area of low mineral concentration in the soil to an area of higher mineral concentration inside the root hair cell) (1) using transport proteins in the cell membrane (1) and energy from respiration. (1)

Unit 2

Page 10

1

Name of tissue	Description	Function
upper and lower epidermis	covers the outer surface of the leaf contains guard cells that form stomata	protection stomata control gas exchange and water loss
layer of palisade cells	box-shaped cells inside the leaf that contain lots of chloroplasts	to absorb light for photosynthesis
air spaces	irregularly shaped cells inside the leaf that form big air spaces	provides a large surface area for gas exchange
xylem	dead, hollow tubes strengthened by lignin	transport water and mineral ions from roots to leaves
phloem	tubes made from long cells with pores on the end walls of each cell	carry cell sap containing dissolved sugar

2 root hair cells; xylem; epidermis; phloem

3 temperature

Page 11

1 thin; transparent; open; light

2 The tissue containing the air spaces has many irregularly shaped cells with gaps between each one. This provides a large surface area to allow gases to diffuse in and out of its cells easily.

3

Physical feature of the palisade layer	Linking word	Explanation of the adaptation
	so	they can efficiently exchange gases.
The cells contain many chloroplasts	because	the cells can be tightly packed together to fit in as many as possible.
They have a box-like shape	to	absorb light for photosynthesis.

Page 12

1 2, 4, 1, 3.

2 More water enters the root hair cells by the process of osmosis.

3 Humidity increased: transpiration rate decreased

Air movement increased: transpiration rate increased; . . . water diffuses out faster.

Light intensity increased: transpiration rate increased; . . . water vapour to diffuse out more easily.

Temperature increased: transpiration rate increased; water evaporates and diffuses out of the leaf faster.

4 When water moves out the guard cells they become flaccid, causing the stoma *to close. This means that much less water vapour can escape from the leaf and the plant is less likely to wilt.*

Page 13

1

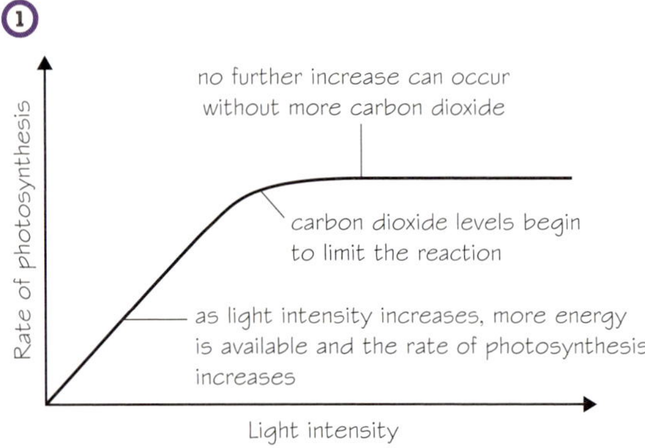

2 Yellow: increases; stays the same; light

Blue: stays the same; increases; carbon dioxide

3

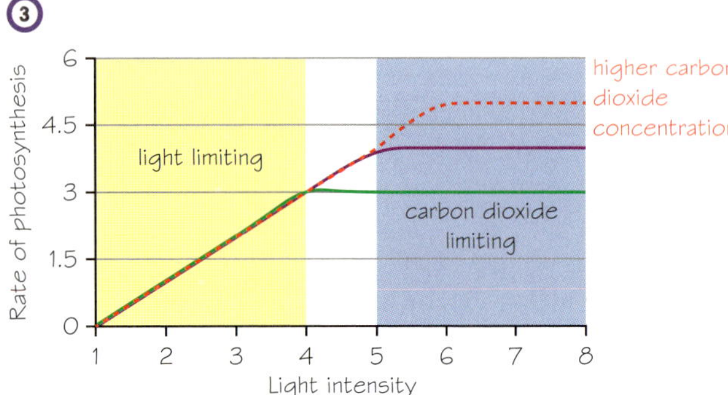

Page 14

1 This light is needed for the chloroplasts to carry out photosynthesis (to produce sugar).

2 The student has described a feature of the layer of palisade cells but not explained how it helps the plant (by absorbing light for photosynthesis).

3 diffuses; stomata; xylem; osmosis

Page 15

1 (a) The closing of guard cells limits the amount of water vapour that can escape from the leaf through the stoma **(1)**, therefore conserving the water inside the plant so the plant is less likely to wilt and die **(1)**.

(b) 20 (arbitrary units)

(c) Any one from: light intensity **(1)**, temperature **(1)**

Page 16

1 (a) It has elongated cells that form tubes allowing it to transport dissolved sugar **(1)**. It has pores/holes on the end wall of each cell to allow the cell sap to move between cells **(1)**.

(b) The movement of water **(1)** from the roots to the leaves **(1)**.

When it is hotter, water particles inside a leaf have more energy and move faster **(1)**, so water evaporates and diffuses out of the leaf faster **(1)**.

Unit 3

Page 18

1 A, c; B, d; C, a; D, b

2 a tt b a non-tongue-roller c Tt

3 a BB b bb c Bb

Page 19

1

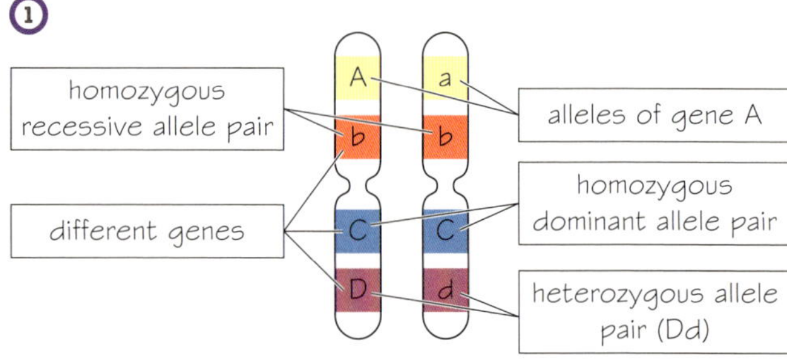

2 Because during fertilisation one chromosome from the pair comes from the female gamete and one chromosome comes from the male gamete.

3 a Bb (or any pair of upper and lower case letters, except Xx and Yy)

b Black fur

4 one chromosome from its father.

codes for fur colour.

alleles.

it will have black fur.

the rabbit will only have white fur if it inherits the recessive allele from both parents.

Page 20

1

	Tall	Short
Parent plant phenotypes	Tall	Short
Parent plant genotypes	TT	tt
Gamete genotypes	T T	t t

2 and 3

	T	T
t	Tt	Tt
t	Tt	Tt

4 Tall

5 All the offspring would be tall because all of the possible offspring would always carry the dominant allele. The dominant allele overrules the recessive allele in a heterozygous genotype.

6 a Tt circled (four)

b For a pea plant to be short it would have two recessive alleles tt. A genotype with a dominant tall allele will always be tall.

Page 21

1 a 4 b 1

c i $\frac{1}{4}$ ii 25% iii 0.25

2

Phenotype	Probability		
	Fraction	Decimal	Percentage
red flowers	$\frac{3}{4}$	0.75	75%

3 a

	r	r
R	Rr	Rr
r	rr	rr

b Rr, Rr c 1:1

d i $\frac{1}{2}$ ii 0.5 iii 50%

Page 22

1 a b b $\frac{1}{4}$

c i 0.25 ii 25%

d A genotype is the type of alleles in an organism and the phenotype means colour of the eyes.

e The phenotype is what an organism looks like. Eye colour is an example of a phenotype.

Page 23

1 (a) a homozygous recessive genotype

(b)

	N	n
N	NN	Nn
n	Nn	nn

(c) 25% or 0.25 or $\frac{1}{4}$

Page 24

1 (a) a variant of a gene (b) Dd (c) dd

(d)

	D	d
d	Dd	dd
d	Dd	dd

50%

Unit 4

Page 26

1 Many bodily processes are controlled by **hormones**. These chemicals are released by **glands** and are transported in the **blood** to their target **organs**. One example is **oestrogen**, which is released by the ovaries and brings about **ovulation**.

2

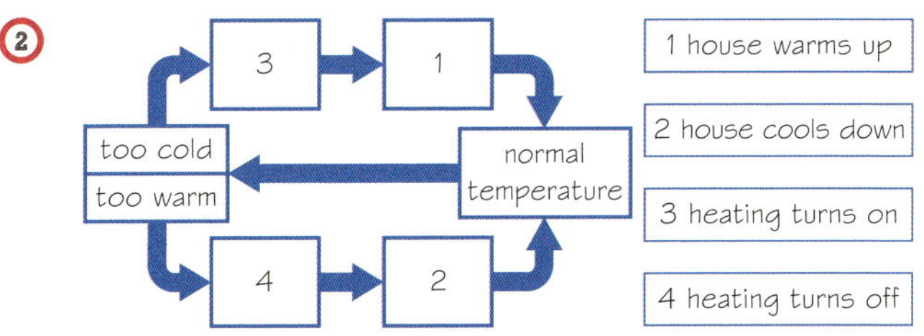

3

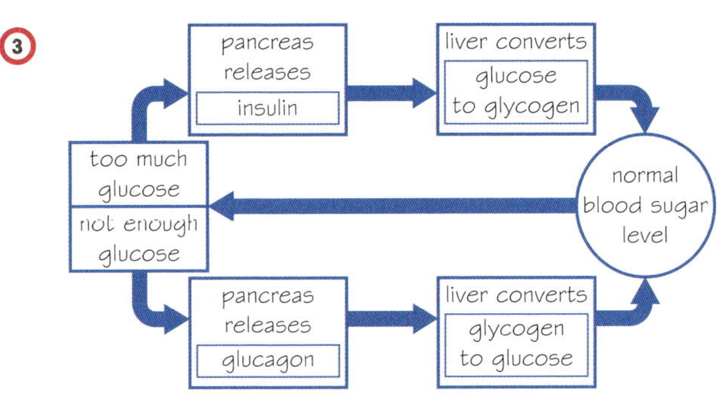

4 ovary; uterus

Page 27

1 From top to bottom: FSH; progesterone; LH; oestrogen

2 4; 8; 10; 9; 1; 6; 7; 5; 2; 3

Page 28

1 a A, c; B, a; C, b

b

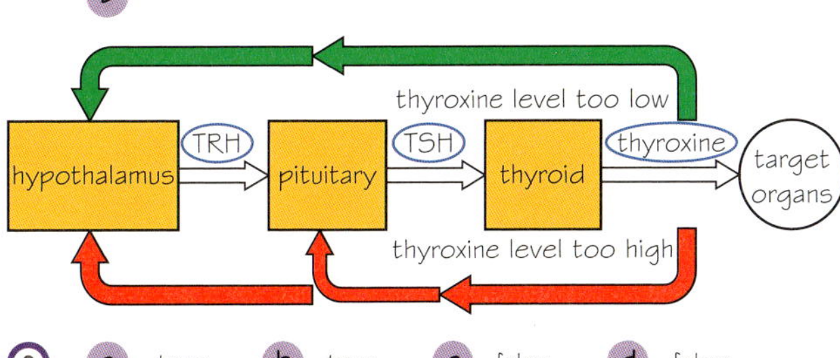

2 a true b true c false d false

Page 29

1 a i decreases / inhibits / reduces it

ii increase

b it is reducing / inhibiting their release

c Normal FSH and LH production will now cause eggs to develop, mature and be released.

2 6; 3; 5; 2; 4; 1

3 a oestrogen b FSH and LH

c The pituitary gland is not producing them. /owtte

4 They collect eggs from the ovary / mix them with sperm / allow 'eggs and sperm put into tube' or 'fertilisation occurs' / fertilised egg then starts dividing / an embryo is formed / embryo put back / inserted into womb / uterus (not vagina).

Page 30

1 a LH b W = FSH

2 a High levels of progesterone inhibit the production of LH by the pituitary gland.

b Reduction in LH levels will lead to reduction of progesterone production / therefore progesterone production is negatively affected by LH levels.

3 Normal levels of FSH / LH will cause many eggs to be matured and released / so increased probability of fertilisation.

Page 31

1 (a) LH **(1)** and FSH **(1)** as both rise / increase (sharply). **(1)**

(b) A reduction in FSH levels will lead to reduction in oestrogen production **(1)** so oestrogen production is negatively affected by high oestrogen levels. **(1)**

(c) A high level of FSH **(1)** is more likely to lead to egg maturity / release / maturation. **(1)**

Page 32

1 (a) As ovulation occurs **(1)** the levels of progesterone released from the corpus luteum increases to maintain the lining of the uterus. **(1)**

(b) Her progesterone level decreases after day 23 to 17.11 **(1)** so uterus wall thickness is not maintained / pregnancy has not occurred / not pregnant. **(1)**

(c) FSH stimulates growth/ maturing of follicle(s)/ eggs/ FSH stimulates oestrogen release / oestrogen stimulates development of uterus lining / oestrogen stimulates LH release / production / LH stimulates ovulation / egg release (any 3, for **(1)** each)

(d) An underactive thyroid would cause less thyroxine to be released **(1)** low levels of thyroxine should stimulate the production of TRH / TSH being released and normally more thyroxine being released **(1)** metabolic rate decreases. **(1)**

Unit 5

Page 34

1 A, d; B, c; C, b; D, a

2 First line: 4; second line: 1; third line: 3; fourth line: 2.

3 B – to add new desirable characteristics

4 To isolate/remove the gene for the desired characteristic.

Page 35

1

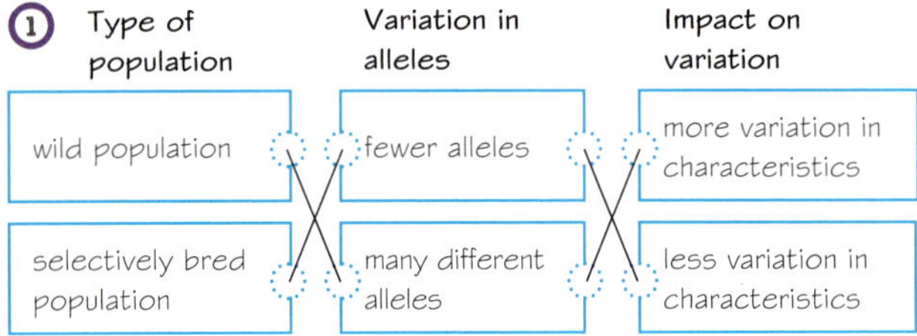

2 C

3 Because there are fewer alleles in selectively bred populations and therefore less chance of the allele offering resistance being present in the population.

4 More; increased

Page 36

1 and 2

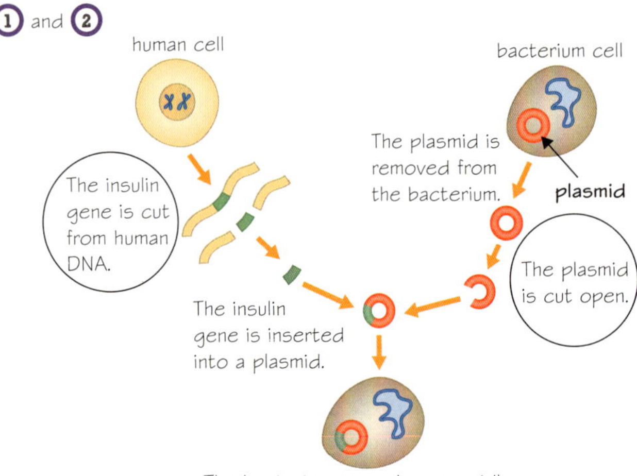

3 the plasmid

4 first row: 3 second row: 1
third row: 4 fourth row: 2

Page 37

1 A, b; B, a; C, c

2 a Food and shelter.

b There would be less food and shelter.

c It would decrease.

d There would be less reproduction of wild plants.

e A f A

3 first row: resistant to herbicide so the crops don't die but the weeds do
second row: light or water
third row: buying and applying herbicide

Page 38

1 a and b

The blood clotting gene is isolated using restriction enzymes that cut the DNA.
The circular bacterial plasmid is then cut open using the same restriction enzymes.

The gene can then be inserted into the plasmid of the bacteria using ligase enzyme.

The bacteria are grown in large tanks.

The bacteria then produce the blood clotting protein which can be extracted.

c A vector is something that can carry a selected gene into a host cell.

Page 39

1 (a) Scientists only breed from the plants with alleles that produce the desirable characteristics **(1)**. Over time, the proportion of alleles for desirable characteristics increases in the population and the plants become more genetically similar to each other **(1)**.

(b) If a new disease affects the wheat crops, the lack of variation in the alleles means if one is affected by it, all the wheat crops are likely to be affected by it **(1)**.

(c) Herbicide/weedkillers can be used without killing the crop **(1)**. Higher yield because there is less competition from weeds **(1)**.

(d) Any **two** from:

- It can have negative effects on populations of wild flowers/insects/reduce biodiversity/affect food chains **(1)**.
- Long-term effects of eating GM crops has not been researched **(1)**.
- Concern over super weeds developing from gene for herbicide resistance being transferred to weeds/ super weeds evolving over time **(1)**.

Page 40

1 (a) The process of transferring a gene for a desired characteristic from one organism's genome into another organism so that it has the desired characteristic **(1)**.

(b) Isolate the genes responsible for producing protein in spider silk using enzymes **(1)**.

Use restriction enzyme to cut plasmid open **(1)**.

Insert the spider silk genes into bacterial plasmid using ligase enzymes **(1)**.

Allow genetically modified bacteria to replicate/grow/ reproduce so they produce the protein **(1)**.

2 (a) Fewer insects to eat crops so higher yield **(1)**. Less effort/work for farmer and therefore more profit/no harmful substances used so better for environment **(1)**.

(b) Insects that don't damage crop may also be killed **(1)**; fewer insect pollinators in environment/damage to food chains/reduction in biodiversity/long-term effects of consuming GM not known/gene could spread to other species **(1)**.

Unit 6

Page 42

1

Standard form	Ordinary number	Standard form	Ordinary number
2×10^5	200 000	6×10^0	6
5×10^4	50 000	1×10^{-1}	0.1
3×10^3	3000	8×10^{-2}	0.08
4×10^2	400	9×10^{-3}	0.009
7×10^1	70	4×10^{-4}	0.0004

2 Top row from left to right: × 1000; × 1000; × 1000
Bottom row from left to right: micrometres; nanometres; picometres

3 a millimetre b metre (m)
c nanometre (nm) d millimetre (mm)

4 a 50 mm; 0.05 mm b $\frac{50\,\text{mm}}{0.05\,\text{mm}}$
c × 1000

Page 43

1 a 9.8 b 5 c 9.8×10^5

2 a 2 b 2 c 2×10^{-2}

3 a $3 \times 10^5 \times 8 \times 10^6$
b 8; 5; 6 24; 6
24; 11 2.4; 12

4 a $\frac{1.2 \times 10^{16}}{3 \times 10^{12}}$
b 3; 16 0.4; 16; 12 0.4; 4 4

Page 44

1 Top row from left to right: × 1000; × 1000; × 1000
Bottom row from left to right: ÷ 1000; ÷ 1000; ÷ 1000

2 a 1000 b 60 000 c smaller

3 a 1 000 000 b 35 000 000 c larger

4 a $\frac{620\,000\,000}{1\,000\,000\,000\,000} = 0.000\,62$
b 0.000 62

5 1; 1000; 1 000 000

6 Chloroplast: 6000
Mitochondrion: 0.4; 400
Ribosome: 0.22; 220
Nucleus: 0.5; 500

Page 45

1 a $\frac{8\,\text{mm}}{40}$; 0.2 b 2.0×10^{-1} c 200 μm

2 a $\frac{9\,\text{mm}}{40}$; 0.225 b 225 μm

3 6×10^{-6} m 4 2.5×10^{-4} mm

Page 46

1 a Divided magnification by image size. Presented incorrect answer in mm not μm.
b 4crf

2 4.2×10^6

3 $\left(\frac{8.5 \times 10^8}{1.7 \times 10^6}\right) = \left(\frac{8.5}{1.7}\right) \times \left(\frac{10^8}{10^6}\right)$
$= 5 \times 10^2$ bacteria per mm^2

Page 47

1 (a) actual size = image size/magnification so $\frac{10}{25} = 0.4$ mm. 1 millimetre = 1000 micrometres so $0.4 \times 1000 = 400$ μm

(b) 3×10^{-3} (correct number between 1 and 10 **(1)**, correct index number 10^{-3} **(1)**)

(c) $(3 \times 4) \times (10^3 \times 10^4) = 12 \times 10^7 = 1.2 \times 10^8$

Page 48

1 $\frac{6}{750} = 0.008\,\text{mm} = 8\,\mu\text{m}$

2 $\frac{2.4 \times 10^9}{5 \times 10^3} = \left(\frac{2.4}{5}\right) \times 10^{9-3}$
$= 0.48 \times 10^6 = 4.8 \times 10^5$

Unit 7

Page 50

1 A, d; B, f; C, e; D, a; E, c; F, b

2 a Evaluate
b C – Give the risks and benefits of IVF

3 6, 2, 4, 3, 5, 1

Page 51

1 a From top to bottom:
Describe, Explain, Compare, Evaluate
b A, c; B, a; C, d; D, b

2 a Evaluate
b Use the information supplied, as well as your own knowledge and understanding, to weigh up the advantages and disadvantages or risks and benefits to come to a judgement.
c genetic engineering
d These genetically modified (GM) tomato plants have resistance to a range of insect pests that non-GM tomato plants do not have. This means that pesticides which pollute the water and soil do not need to be applied.

Page 52

1 a The stem cells can develop into most other types of cell.
Each stem cell divides every 30 minutes.
There is a low chance of a patient's immune system rejecting the cells.
b It costs £5000 to collect a few cells.
There are ethical issues in using embryo stem cells.
More research is needed into the use of these stem cells.
c It costs £1000 to collect many cells.
Adults give permission for their own bone marrow to be collected.
Use of these stem cells is considered to be a safe procedure.
d The stem cells can develop into only a few types of cell.
Each stem cell divides every 4 hours.
There is a high chance of a patient's immune system rejecting the cells.

2 The benefits of using embryonic stems cells are that: they divide into other types of cells / the cells divide quickly / there is no rejection.

The risks of using embryonic stem cells are that: they are expensive / there are ethical considerations / more research is needed.

The benefits of using bone marrow stem cells are that: adults give permission / it is a safe procedure / it is relatively inexpensive.

The risks of using bone marrow stem cells are that: there are limited types of cell / the cells divide slowly / there is a higher rejection risk.

3 embryonic; bone marrow (either way round)

If embryonic stem cells preferred: the benefits of the faster cell growth rate with low rejection rates and a greater range of cell types than with bone marrow stem cells outweigh the risks more than for bone marrow stem cells.

If bone marrow stem cells preferred: the benefits of adults being able to give permission, the procedure being considered to be safe and the procedure being cheaper than using embryonic stem cells outweigh the risks more than for embryonic stem cells.

Page 53

1 irrelevant; for; against; for; irrelevant

2 a 5.5%; 5.8% b £100; £112

c From the table it appeared that drug B was the best. However, I think that drug A is the most effective because it is more cost effective per patient. It costs £100 per patient with a 5.5% failure rate compared with drug B, which costs £112 per patient with a 5.8% failure rate.

Page 54

1 No

2 *As the BMI goes up, the number of cases of type 2 diabetes also goes up. Men have a greater risk of developing type 2 diabetes than women because the line of their graph is higher than that of women.*

3 *BMI means Body Mass Index and over 30 means that a person is obese.*

4 *At greater than 40 BMI, there were 58 new cases per 1000 women per year. At less than 20 BMI, there were 20 new cases per 1000 men per year.*

5 Compare the risk for men/women using the data at different BMIs. Justify the higher incidence in men at lower BMI values by using data from the graph.

6 No

7 Compare the number of cases of type 2 diabetes in men with the number of cases in women at different BMIs with supporting evidence. For example, BMI less than 20 is 20 cases in men compared with 8 in women, which is 2.5 times higher; at BMI 30–35 it is 42 cases for men compared with 28 for women, which is 1.5 times more; at BMI greater than 40 BMI it is 70 cases for men compared with 58 for women, which is 1.2 times more. Finish the answer with a concluding sentence, for example, the risk of type 2 diabetes is greater for both men and women at high BMI. For men, the risk of type 2 diabetes is greater than for women at low BMI (less than 20), but the difference in the risk between men and women is much lower at high BMI (greater than 40).

Page 55

1 Evaluate

2

		Patient numbers two years after artificial hearts implanted	
Design of heart	**Total number of patients in study**	**% of patients who are still alive**	**% of patients who died**
A: new design	90	$\left(\frac{63}{90}\right) \times 100 = 70$	$\left(\frac{27}{92}\right) \times 100 = 30$
B: old design	150	$\left(\frac{72}{150}\right) \times 100 = 48$	$\left(\frac{78}{150}\right) \times 100 = 52$

3 7, 4, 8, 6, 3, 5, 1, 2

4 The new design was tested on a smaller sample and 63 patients survived. The old design was tested on a larger sample and 72 patients survived. The survival success rate of the new design after two years is 70%. The survival success rate of the old design after two years is 48%. Even though the new design was tested on a smaller sample and a lower number of people survived, the actual survival rate as a percentage was much higher than that of the older design. So, the new design is the one I would choose to implement.

Page 56

1 Answer could include the following points in a logical order for 6 marks:

IVF has made it possible for the woman to have the opportunity to have a child at 55 when it was previously not possible. This may also mean she has more time and money available to provide good parenting. The table shows that there was a 20% success rate for women in her age range at the IVF clinic she attended compared with 40% at age 30–39. However, there is still the possibility of a successful pregnancy.

The downside is that having a child at 55 will mean the woman will be quite elderly when the child reaches adulthood and there are more risks associated with giving birth at 55. Information from the table shows that there is an increased risk of multiple births with increasing age. From an average of 2.1 embryos transferred at 30–39 years up to 3.1 embryos on average transferred at 50–59 years old range. So, even though the treatment is less successful as a woman increases in age, there is a greater risk of having multiple births. This may not be what the woman wants.

I think that she should not have gone ahead with IVF treatment. An increased chance of an unsuccessful pregnancy would cause a great deal of stress and anxiety. If successful, the increased possibility of multiple births within her age range means there could be several children which may become very difficult to manage as she gets older.

Chemistry

Unit 1

Page 58

1 $C_3H_8 + 5O_2 \rightarrow 3CO_2 + 4H_2O$

2 A, b; B, a; C, b; D, b

3 7 mol of atoms (1 N atom, 5 H atoms, 1 O atom), 1 mol of NH_4^+ ion, 1 mol of OH^- ion

4 This reaction has 1 mol of Cu^{2+} ions, 2 mol of electrons and 1 mol of Cu atoms.

Page 59

1 7

2 The relative particle mass could be:

- the relative **atomic** mass, A_r, found in the **periodic** table
- the **relative formula mass**, M_r, calculated from A_r values and the chemical formula.

3 32 + 16 + 16 = 64

So 1 mol of SO_2 molecules has a mass of 64 g.

4

Substance	Particle type	Formula	Relative particle mass	Mass of 1 mol (g)
Sodium chloride	ions	NaCl	23 + 35.5 = 58.5	58.5
Magnesium	atom	Mg	24	24
Water	molecule	H_2O	(2 x 1) + 16 = 18	18
Glucose	molecule	$C_6H_{12}O_6$	(6 × 12) + (12 × 1) + (6 × 16) = 180	180
Sulfate	ion	SO_4^{2-}	32 + (4 × 16) = 96	96
Potassium	atom	K	39	39

Page 60

1 a 185.25 g

b 1 Cu atom, 1 C atom, 3 O atoms so M_r = 63.5 + 12 + (3 × 16) = 123.5

c $= \frac{185.25}{123.5} = 1.5$ mol

2 213 g and 71

number of moles $= \frac{213}{71} = 3$ mol

3 3 mol × 6.02 × 10^{23} = 1.806 x 10^{24} = 1.81 × 10^{24} molecules (3 sf)

Page 61

1 a 224 g b 16 g

c

	Iron (Fe)	Oxygen (O)
Mass (g)	224	96
Relative atomic mass, A_r	56	16
$\frac{\text{mass}}{A_r}$ = number of moles	$\frac{224}{56} = 4$	$\frac{96}{16} = 6$
Find the simplest ratio of the number of moles by dividing by the smallest number	$\frac{4}{4} = 1$	$\frac{6}{4} = 1.5$
If needed, multiply by 2 to make the simplest ratio as whole numbers	2	3
Ratio of atoms	2 iron atoms for every 3 oxygen atoms	
Empirical formula	Fe_2O_3	

2 a 5.4 g and 21.3 g b The periodic table

c

	Aluminium (Al)	Chlorine (Cl)
Mass (g)	5.4	21.3
Relative atomic mass, A_r	27	35.5
$\frac{\text{mass}}{A_r}$ = number of moles	$\frac{5.4}{27} = 0.2$	$\frac{21.3}{35.5} = 0.6$
Find the simplest ratio of the number of moles by dividing by the smallest number	$\frac{0.2}{0.2} = 1$	$\frac{0.6}{0.2} = 3$
Ratio of atoms	1 aluminium atom for every 3 chlorine atoms	
Empirical formula	$AlCl_3$	

Page 62

1 a highlighted: 12, formula, 69, mol

b circled: 23, 16 , 51, 51, 1.3529, 1.4

c M_r = (2 × 23) + 12 + (3 × 16) = 106; $\frac{69}{106}$ = 0.65 mol

2 a First row: masses of lead and bromine have been swapped in error. M_r of lead stated as 270 when it is 207.

b

Lead (Pb)	Bromine (Br)
$\frac{41.4}{207} = 0.2$	$\frac{32}{80} = 0.4$
$\frac{0.2}{0.2} = 1$	$\frac{0.4}{0.2} = 2$
Empirical formula is $PbBr_2$	

Page 63

1 (a) M_r of $CaCO_3$ = 40 + 12 + (3 × 16) = 100 (1) or 100 g = 1 mol; number of moles $= \frac{286}{100}$ = 2.86 mol (1)

(b) 5 × 6.02 × 10^{23} = 5 × 6.02 × 10^{23} = 3.01 × 10^{24} molecules (1)

(c)

	Ca	O	
No. of moles	$\frac{50}{40} = 1.25$	$\frac{20}{16} = 1.25$	(1)
Ratio	1	1	(1)
Empirical formula	CaO		(1)

Empirical formula = CaO

Page 64

1 M_r = (2 × 7) + 32 + (16 × 4) = 110 (1)

Number of moles of lithium sulfate $= \frac{282}{110}$ = 2.56 mol (1)

Number of moles of Li ions = 5.12 mol (1)

2 0.5 × 6.02 × 10^{23} = 3.01 × 10^{23} (1)

3

C	H	Cl	
$\frac{360}{12} = 30$	$\frac{60}{1} = 60$	$\frac{1065}{35.5} = 30$	(1)
$\frac{30}{30} = 1$	$\frac{60}{30} = 2$	$\frac{30}{30} = 1$	(1)
Empirical formula CH_2Cl			(1)

Unit 2

Page 66

1 mass of solute (g) = concentration (g dm^{-3}) × volume of solution (dm^3)

volume of solution (dm^3) $= \frac{\text{mass of solute (g)}}{\text{concentration (g dm}^{-3})}$

2 A_r or $M_r = \frac{\text{mass of substance}}{\text{number of moles}}$

3 number of moles $= 0.6 = \frac{\text{mass}}{23}$

$0.6 \times 23 = \left(\frac{\text{mass}}{23}\right) \times 23$

mass = 13.8 g

Page 67

(1) 3, 6, 4, 2, 1, 5

(2)
1. 0.03 highlighted
2. mass = number of moles × M_r
3. 12 + (4 × 1) = 16
4. 0.03 × 16
5. 0.48 g

(3)
1. 0.31 kg highlighted
2. number of moles = $\frac{\text{mass (g)}}{M_r}$
3. (2 × 23) + 16 = 62

   0.31 kg = 310 g
4. $\frac{310\text{ g}}{62}$
5. 5.0 mol

Page 68

(1)
a. Highlighted as follows: 0.0801, 24.9
b. Highlighted as follows: 0.0801, 24.9
c. 37.4 has 3 significant figures: 3, 7 and 4
d. If you write 37.4 to 1 significant figure, then the first sf is 3. The next digit is 7, so you round 3 up to 4. You need to write a zero to keep the place value. So, 37.4 to 1 sf is 40.

   If you write 37.4 to 2 sf, then the second sf is 7. The next digit is 4, so do not round up. 37.4 to 2 sf is 37.

(2) a 3 b 2 c 4 d 2 e 1 f 3

(4)

Number	To 1 sf	To 2 sf	To 3 sf
0.02564	0.03	0.026	0.0256
0.000 839 21	0.0008	0.000 84	0.000 839
1.035	1	1.0	1.04
609.72	600	610	610

Page 69

(1)
a. $CuSO_4$ and $MgSO_4$ highlighted
b. 63.5 + 32 + (4 × 16) = 159.5

(2)
a. $\frac{319}{159.5}$ = 2 mol
b. 24 + 32 + (4 × 16) = 120
c. 1 mol of $CuSO_4$ produces 1 mol of $MgSO_4$. This gives a mole ratio of 1 : 1.
d. 2 mol
e. 2 × 120 = 240 g (2 sf)

(3)

Substance	$CuSO_4$	$MgSO_4$
Mass (g)	319	240
M_r	159.5	120
Mole ratio	1	1
Number of moles	2	2

Page 70

1 mass = moles × M_r

M_r of $Pb(NO_3)_2$ = 207 + 2 × (14 +(3 × 16)) = 207 + (2 × 62) = 331

mass = 1.68 × 331 = 556 g (3 sf)

Page 71

1 (a) M_r H_2SO_4 = (2 × 1) + 32 + (4 × 16) = 98

(b) M_r NH_3 = 14 + (3 × 1) = 17

number of moles of NH_3 = $\frac{\text{mass}}{M_r} = \frac{3.4}{17}$ = 0.2 mol

From the balanced equation, 2 mol NH_3 produces 1 mol $(NH_4)_2SO_4$, so the mole ratio is 2 : 1.

0.2 mol NH_3 produces $\frac{0.2}{2}$ = 0.1 mol $(NH_4)_2SO_4$

mass of $(NH_4)_2SO_4$ produced = moles × M_r

M_r $(NH_4)_2SO_4$ = (2 × 14) + (2 × 4 × 1) + 32 + (4 × 16) = 132

mass = 0.1 × 132 = 13.2 g = 13 g (2 sf)

Page 72

1 M_r $CaCO_3$ = 40 + 12 + (3 × 16) = 100

Number of moles of $CaCO_3$ = $\frac{115}{100}$ = 1.15 mol

From the balanced equation, the mole ratio is 1 : 1,

so 1.15 mol $CaCO_3$ produces 1.15 mol $Ca(NO_3)_2$

M_r $Ca(NO_3)_2$ = 40 + (2 × 14) + (2 × 3 × 16) = 164

mass = moles × M_r = 1.15 × 164 = 188.6 g = 189 g (3 sf)

Unit 3

Page 74

(1) CO_2

(2) a Al: 1 H: 3 b S: 2 O: 8

(3) (s); (l); (g); (aq) (4) A_r or M_r (5) SO_4(aq)

Page 75

(1)
a.

Element or group	Number of each atom, or group of atoms	
	Reactants	Products
Al	2	2
O	7	13
H	2	2
S	1	3
SO_4	1	3

b. reactants, More, reactants c. 3

d.

Element or group	Number of each atom, or group of atoms	
	Reactants	Products
Al	2	2
O	15	13
H	6	2
S	3	3
SO_4	3	3

e. oxygen/O and hydrogen/H f. 2 O, 4 H

g. water/H_2O h. 3 i. 3;3

Page 76

(1)
a.

Li	O_2
56	64
7	32
$\frac{56}{7}$ = 8	$\frac{64}{32}$ = 2
4	1

b. 4; 1

(2) a 4; 2 b 2; 1 c 4; 2

Page 77

1. a Na^+; Cl^- b $H^+(aq)$; $OH^-(aq)$
2. a K^+; Cl^-; Ag^+; NO_3^- b K^+; NO_3^-
   c $Ag^+(aq)$; $Cl^-(aq)$; $AgCl(s)$
3. 2; 2

Page 78

1. a sulfuric acid / H_2SO_4; lithium hydroxide / LiOH; lithium sulfate / Li_2SO_4; water
   b no c no
   d $H_2SO_4 + LiOH \rightarrow Li_2SO_4 + H_2O$
   e

Element	Number of each atom	
	Reactants	Products
H	3	2
S	1	1
O	5	5
Li	1	2

   f S; O g H; Li h both
   i $H_2SO_4 + 2LiOH \rightarrow Li_2SO_4 + 2H_2O$
   j

Element	Number of each atom	
	Reactants	Products
H	4	4
S	1	1
O	6	6
Li	2	2

Page 79

1 $Na_2CO_3 + 2HNO_3 \rightarrow 2NaNO_3 + H_2O + CO_2$

all formulae on correct side **(2)**

balancing **(1)**

allow $\frac{3}{4}$ formulae **(1)**

2

	MgO	HNO_3
mass (g)	120	378
M_r	40	63
number of moles	$\frac{120}{340} = 3$	$\frac{378}{63} = 6$
simplest mole ratio	1	2

$MgO + 2HNO_3 \rightarrow Mg(NO_3)_2 + H_2O$

calculated mol of MgO **(1)**

calculated mol of HNO_3 **(1)**

determined simplest ratio / LHS of equation **(1)**

deduced formula of $Mg(NO_3)_2$ produced / RHS of equation **(1)**

Page 80

1 LHS: 14 **(1)**

RHS: 9; 10 both numbers correct **(1)**

2 $Mg^{2+}(aq) + 2OH^-(aq) \rightarrow Mg(OH)_2(s)$ **(1)**

balancing of correct formulae **(1)**

correct state symbols: $Mg^{2+}(aq)$; $2OH^-(aq)$; $Mg(OH)_2(s)$ **(1)**

allow reactants in any order

Unit 4

Page 82

1. a ammonium chloride → hydrogen chloride + ammonia
   b ammonium chloride
   c hydrogen chloride; ammonia
   d hydrogen chloride; ammonia
   e ammonium chloride f the same
2. A conical flask with a tight-fitting bung. A bottle with a screw top.
3. A, b, e; B, a, c
4. a C b B

Page 83

1. a, b, c, d

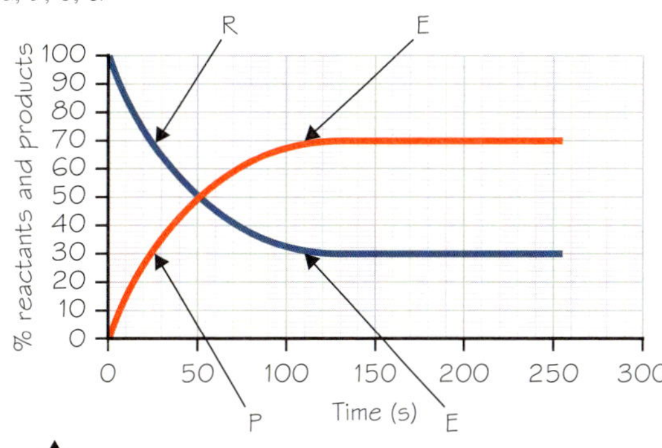

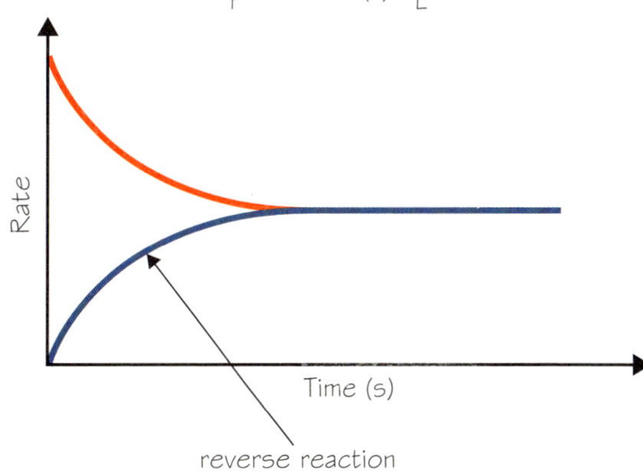

2. remains the same; equal to

Page 84

1. a forward reaction = exothermic; reverse reaction = endothermic
   b 2nd row: Decreases 3rd row: Increases
   c exothermic; Increasing
2. 2nd row: Moves to the left
   3rd row: Decrease; The reverse reaction is endothermic and the reaction changes to increase the temperature.

Page 85

1. a Produce more ●
   b Produce less ● + ●
   Shift the position of equilibrium towards the right
2. four/4; two/2; right; ammonia/NH_3

Page 86

1. 1 B has specified the actual reactions with equal rates. The rate of the forward reaction is equal to the rate of the reverse reaction. A hasn't said this.
   2 B has correctly stated the required conditions but A has not mentioned that there must be a closed system.

2 Both C and D got a mark for 'endothermic' but D did not get the reason right and C did.

Page 87

1 (a) When the forwards and backwards reactions in a reversible chemical reaction are occurring at the same rate **(1)** in a closed system. **(1)**

(b) The brown colour becomes paler. **(1)**

(c) More NO_2 is produced **(1)** as the position of equilibrium moves to the right **(1)** because the forward reaction is endothermic. **(1)**

Page 88

1 (a) They are the same **(1)** because the system is at dynamic equilibrium **(1)**.

(b) The equilibrium is lost **(1)** because gas escapes from an open system **(1)**.

2 (a) Three molecules of gas are reduced to two molecules of gas during the forward reaction. **(1)** Increasing the pressure favours the forward reaction. **(1)**

(b) The forward reaction is exothermic and favours lower temperatures. **(1)** At higher temperatures, less SO_3 would be produced/the reverse reaction is favoured. **(1)**

Unit 5

Page 90

1 Endothermic: Ice pack; Thermal decomposition of copper carbonate

Exothermic: Hand warmers; Neutralisation reactions; Burning fuels

2 Reaction 1: 5

Reaction 2: −3; endothermic

Reaction 3: −15; endothermic

Reaction 4: 21; exothermic

3 Chemical; collide; minimum; activation

Page 91

1 b, c, e

2 a

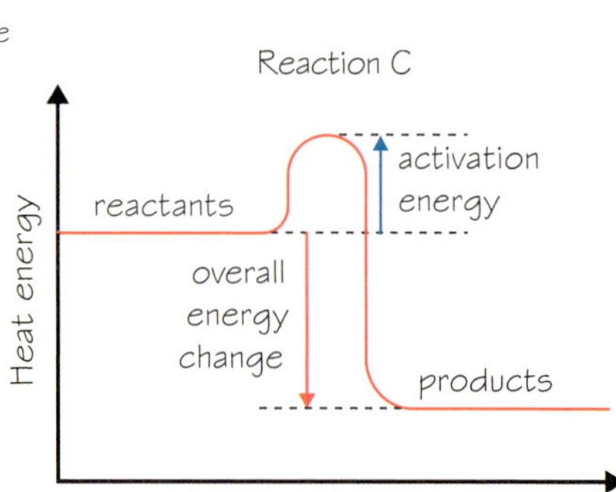

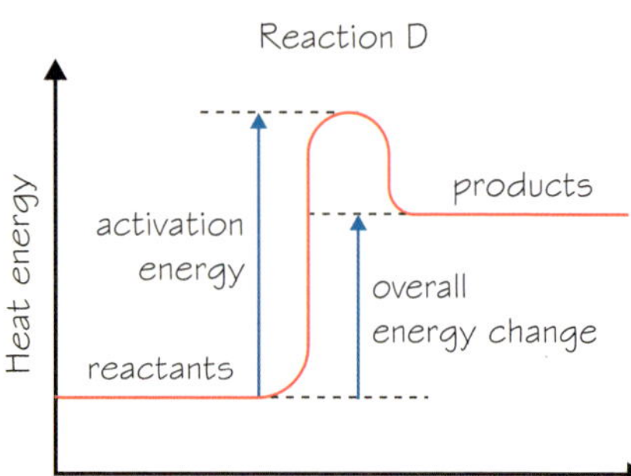

b Reaction C is exothermic because energy is **given out to** the surroundings.

Reaction D is endothermic because **energy is taken in from the surroundings**.

Page 92

1 a B, E

b, c, d

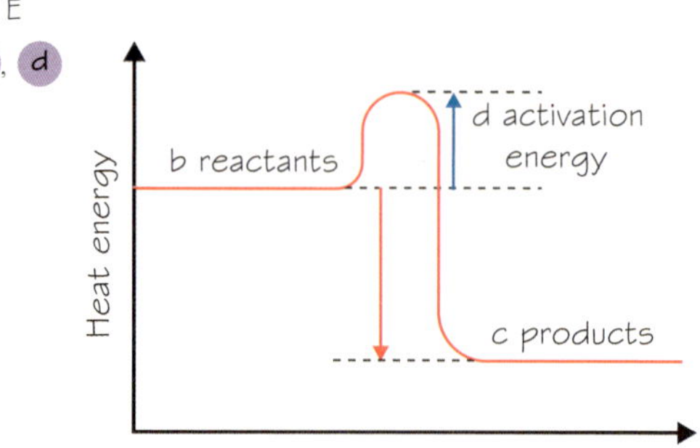

2 endothermic; decreased; less

3 a, b, c, d, e

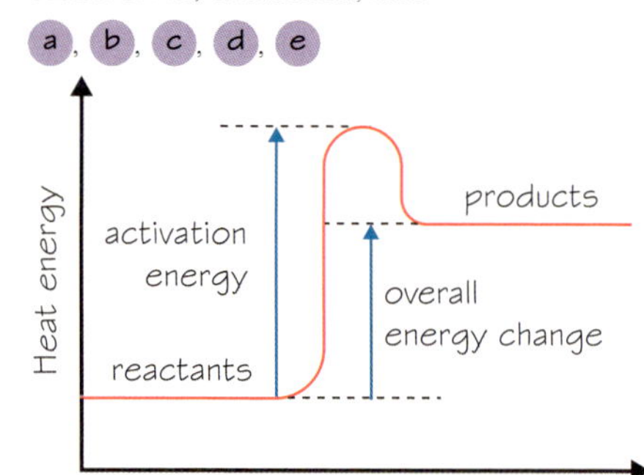

Page 93

1 endothermic; exothermic; bond

2 a 2

b **Step 1:**

$1 \times$ H–H $= 436$ kJ mol^{-1}

$1 \times$ Cl–Cl $= 243$ kJ mol^{-1}

Total energy in $= 679$ kJ mol^{-1}

c **Step 2:**

$2 \times$ H–Cl $= 2 \times 432$ kJ mol^{-1}

Total energy out $= 864$ kJ mol^{-1}

d **Step 3:**

Energy change = energy in − energy out $= 679 - 864 = -185$ kJ mol^{-1}

e It shows that the energy out is greater than the energy in.

f Exothermic

Page 94

1 Two H–H bonds are broken but only one was used in the calculation; four O–H bonds are made but only two were used in the calculation.

2 By being more specific and saying energy is given out when making bonds and taken in when breaking bonds.

Page 95

1 (a) The relative amount of energy stored in the products is less than that stored in the reactants **(1)**. Therefore energy is given out to the surroundings **(1)**, making the reaction exothermic.

(b) Calculate energy needed to break bonds **(1)**

Bonds broken $= 4 \times$ (C–H) $+ 2 \times$ (O=O)

Energy in $= (4 \times 413) + 2 \times$ (O=O) $= 1652 + 2 \times$ (O=O) kJ mol^{-1}

Calculate energy released in forming bonds **(1)**

Bonds made $= 2 \times$ (C=O) $+ 4 \times$ (O–H)

Energy out $= (2 \times 805) + (4 \times 464) = 1610 + 1856 = 3466$ kJ mol^{-1}

Calculate energy change **(1)**

Energy change = energy in − energy out

$-818 = 1652 + 2 \times (O{=}O) - 3466$

$-818 - 1652 + 3466 = 2 \times (O{=}O)$

Evaluation of final answer **(1)**

$2 \times (O{=}O) = 996\,kJ\,mol^{-1}$

$O{=}O = \frac{996}{2} = 498\,kJ\,mol^{-1}$

or

Energy in = 1652 + 2x

Energy out = 3466

Energy change = energy in − energy out

$-818 = 1652 + 2x - 3466$

$-818 = -1814 + 2x$

Add 1814 to both sides

$-818 + 1814 = 2x$

$996 = 2x$

Divide both sides by 2

$498 = x$

Page 96

1 (a) At the start of the reaction, the energy level increases to a maximum and then decreases rapidly **(1)**. The final energy level is lower than at the start of the reaction **(1)**.

(b) Calculate energy needed to break bonds **(1)**

Bonds broken = 1 × (N≡N) + 3 × (H–H)

Energy in = 945 + (3 × 436) = $2253\,kJ\,mol^{-1}$

Calculate energy released in forming bonds **(1)**

Bonds made = 6 × (N–H)

Energy out = (6 × 391) = $2346\,kJ\,mol^{-1}$

Calculate energy change **(1)**

Energy change = energy in − energy out

2253 − 2346 =

Evaluation of final answer **(1)**

$-93\,kJ\,mol^{-1}$

Unit 6

Page 98

1 A, c; B, b; C, a; D, f; E, g; F, d; G, e

 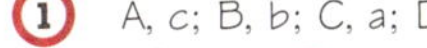 

2 a and b

A An investigation into the effect of different catalysts on the time taken to produce $10\,cm^3$ of oxygen during the decomposition of hydrogen peroxide.

B The effect of changing acid concentration on the temperature changes that take place during a neutralisation reaction.

C The effect of different mobile phases on the position of spots seen on a chromatogram.

3 cannot; small; one data point; not possible

Page 99

1 a Increasing the current during the electrolysis will increase the mass of metal deposited on the negative electrode.

b Make up a solution of copper sulfate or copper chloride solution to test.

In electrolysis, electricity is used to split up compounds, so we need to set up an electric circuit.

Ammeters are used to measure current. Graphite electrodes are inert.

c

Independent variable	current
Dependent variable	mass of metal deposited at the negative electrode
Control variables	salt solution / concentration of the salt solution / graphite electrodes / time circuit switched on

Page 100

1 a i Independent variables: A – electrical current; B – electrodes

ii Dependent variables: A – mass of copper; B – products produced at the cathode

iii Control variables:

A – copper salt, type of electrode, concentration of salt solution

B – copper salt, electrical current, concentration of salt solution

b i 0.1–0.5 A

ii 0.2 A and 1 A are not ranges. 0.1–0.5 A is big enough to show a pattern. The values of 0.01 A and 0.02 A are too close together.

Page 101

1 a A b C c B d D

2 a Mass of cathode at end 2.99 g and/or Mass increase 0.63 g

b Maybe the current was not controlled properly and it went higher / misread the number on the balance.

c Each individual measurement / data point will have some errors in it. By taking the mean, some of the errors will cancel out.

d reproducible; low; one decimal place; precise

Page 102

1 a We don't really know because it is hard to see the pattern when there are no repeats. They could compare their results with those collected by student B.

b Repeat each experiment and take the mean.

c It is an anomalous result and did not fit with the other two data points, which were exactly the same.

d The control variables may not be properly controlled or an error is made when reading or recording any data being measured.

Page 103

1 (a) chlorine gas **(1)**

(b) Place some damp litmus paper on / near the anode. **(1)** If it is bleached and goes white, chlorine gas is a product. **(1)** (Note: credit correct test given for a wrong answer, e.g. oxygen in part (a))

(c) (i) $Cu^{2+} + 2e^- \rightarrow Cu$ **(1)**

(ii) mass of the cathode at the start and finish of each experiment / change in mass of cathode **(1)**; the current **(1)**

(iii) Any **two** from: concentration of the copper chloride solution **(1)**; the time the current is switched on **(1)**; voltage during each experiment **(1)**; the graphite electrodes **(1)**

(iv) Repeat each experiment two / three times **(1)** and take the mean results **(1)**.

Page 104

1 Wear eye protection.

Mix the sodium carbonate solution and the zinc carbonate solution in a beaker **(1)**, then filter the mixture **(1)**.

To make sure it is pure, rinse the beaker with a little distilled water and pour the distilled water over the precipitate in the funnel **(1)**.

Carefully remove the filter paper containing the precipitate and dry it in a warm oven **(1)**.

2 potassium hydroxide + sulfuric acid → potassium sulfate + water

$$2KOH + H_2SO_4 \rightarrow K_2SO_4 + 2H_2O$$

Wearing safety glasses, fill a burette with dilute sulfuric acid ensuring the meniscus is at 0 cm^3. Measure out 25 cm^3 of potassium hydroxide solution into a conical flask using a pipette and pipette filler. **(1)**

Add a few drops of an indicator (e.g. phenolphthalein is pink in alkaline solutions and turns colourless at the end point). **(1)**

Add the acid slowly from the burette, swirling the conical flask to mix the acid and alkali thoroughly. Note the volume of acid required for the colour change. **(1)**

Repeat the experiment but this time add the acid drop by drop near the end point to improve the accuracy. **(1)**

To produce the salt crystals, repeat the experiment without the indicator, adding the exact volume of acid to the alkali. **(1)**

Evaporate the water from the solution to produce potassium sulfate crystals. **(1)**

Unit 7

Page 106

1 Link your ideas together to show understanding.

Use correct scientific vocabulary.

Write your answer in an ordered way.

2 A, c, e; B, b, d; C, a, f

Page 107

1
- a Describe
- b Write down some key facts from the information provided.

2
- a relative; based on
- b The word 'relative' is important because it is telling you that you must compare the reactivity of the different metals.

  'Based on' means you use the facts in the table.
- c experimental data

3
- a What you see/observe when different metals react with different substances.
- b Reaction with water; reaction with hydrochloric acid.
- c copper; lithium; zinc; magnesium
- d i copper = zinc, magnesium, lithium

  ii copper, zinc, magnesium, lithium
- e Both copper and zinc do not react with water, but zinc does react with hydrochloric acid. This tells me that zinc is more reactive than copper.

Page 108

1 These questions are **not** relevant:

Consider the variables involved in the reaction. What needs to be changed or controlled?

How will I make it a fair test?

How will I record your results?

2
- a reactants, products and states
- b What is the chemical reaction?

  What reactants will I need?

  What equipment will I use?

  What will I observe or measure?

Page 109

1 When you explain something you must give a reason for your answer. Reasons are not required when you describe something.

2 relative reactivity; halogens; electronic configuration

3 The reactivity of halogens decreases down the group.

The electronic configuration is the way in which an atom's electrons are arranged.

The electronic configuration of chlorine is 2, 8, 7.

As you go down the group, the distance between the electron in the outermost shell and the nucleus increases.

The formation of negative ions is important for reactivity.

The force of attraction depends on distance from the nucleus.

4 The relative reactivity of the halogens – most to least reactive: fluorine, chlorine, bromine, iodine.

Page 110

1
- a Describe
- b first paragraph
- c No
- d No

2 Take two same-sized pieces of the copper and place each into a test tube. Add 5 cm^3 of zinc sulfate to one and 5 cm^3 of magnesium sulfate to the other. Record any observations. Repeat the same process with pieces of zinc added to copper sulfate and magnesium sulfate. Then repeat with pieces of magnesium added to copper sulfate and zinc sulfate.

Record the observations in this table. If there is a reaction you may see a colour change, the metal disappearing and a new metal deposited. An increase in temperature could be measured as the reaction is exothermic.

The most reactive metal is the one that reacts with both sulfate solutions and the least reactive metal is the one that reacts with neither.

	Observations		
	copper sulfate	zinc sulfate	magnesium sulfate
Copper	X		
Zinc		X	
Magnesium			X

Page 111

1
- a Describe
- b Give an account of something. Statements need to be developed as they are often linked. You do not need to give reasons.
- c hydrochloric acid
- d Heating the reactants: D; Filtering: C; Evaporation: A; Crystallisation: B
- e 4; 2; 3; 6; 1; 5

2 Answer could include the following points in a logical order for 6 marks:

Accept either equation as both will produce the same result.

copper oxide + hydrochloric acid → copper chloride + water

$CuO(s) + 2HCl(aq) \rightarrow CuCl_2(aq) + H_2O(l)$

or

copper carbonate + hydrochloric acid → copper chloride + water + carbon dioxide

$CuCO_3(s) + 2HCl(aq) \rightarrow CuCl_2(aq) + H_2O(l) + CO_2(g)$

Wearing safety glasses, measure out some hydrochloric acid in a measuring cylinder and transfer it to a beaker. Add some copper oxide / copper carbonate to the acid and mix well. Heat gently until no more solid dissolves.

Fold a piece of filter paper and place it in a filter funnel in the top of a conical flask or beaker. Filter the hot mixture through the funnel. Pour the filtrate into an evaporating basin and gently warm over a water bath until the volume has reduced by about half. Remove the evaporating basin from the heat and leave the concentrated solution to cool and the copper chloride crystals to form. Finally dry the crystals by dabbing them on a piece of filter paper.

Page 112

1 Answer could include the following points in a logical order for 6 marks:

Diamond and graphite are allotropes of carbon. They are both covalent, giant molecular structures with very high melting points. This is because of the strong covalent bonds between the carbon atoms which need to be broken to change them from a solid to a liquid.

However, the structural arrangements of the carbon atoms are different, which explains the differences in their properties and uses.

Diamond has a tetrahedral arrangement of carbon atoms, which makes it very hard and rigid. It is used as a drilling and cutting tool because of this. This shape also means that diamond can be cut with very flat sides that reflect light so it sparkles, which can be used in jewellery.

Graphite is softer than diamond because it has a layered structure. Each carbon atom is covalently bonded to three others forming layers containing hexagonal rings. There are weak intermolecular forces between the layers, which allow them to slide over each other. This makes graphite a useful lubricant for moving parts. There are also delocalised electrons, which are free to move as they are not all held in the covalent bonds like diamond. This means that graphite conducts electricity but diamond does not. Graphite can be used as electrodes for electrolysis, which is useful as it is not very reactive.

2 Answer could include the following points in a logical order for 6 marks:

copper carbonate + hydrochloric acid → copper chloride + water + carbon dioxide.

As carbon dioxide is given off, the reaction can be followed by measuring the volume of gas produced using a gas syringe attached to the top of a conical flask.

Wear safety glasses throughout the experiment and wipe up any spillages immediately.

For each experiment, use the same:

- mass of copper carbonate (2 g measured on a balance);
- volume of dilute hydrochloric acid (50 cm^3 measured in a measuring cylinder);
- time to collect the carbon dioxide gas (30 seconds measured using a stop watch).

Temperature is the only variable to be changed. This can be done using a large beaker containing five different temperatures of water to get a good range of results. Ice can be used to cool the water bath below room temperature. Different amounts of water from a boiled kettle can be used for the higher temperatures. (Take care with hot water.)

Put a conical flask containing 50 cm^3 dilute HCl into the water bath to give it time to warm up or cool down. Record the temperature of the acid. Add the copper carbonate and seal the gas syringe onto the top of the conical flask. Start timing and measure the volume of gas collected after 30 seconds.

Record the results in a table with these headings.

Temperature of hydrochloric acid in °C	Volume of gas produced in 30 seconds in cm^3

Ideally the whole experiment should be repeated to confirm the trend. A graph of volume of carbon dioxide produced (cm^3) against temperature can be drawn to show the effect of temperature on this reaction.

Physics

Unit 1

Page 114

1 a Work done by a force. b Heating.
c Work done by an electrical current.

2 an electric current; a force; heating

3 From top: gravitational store of the object; elastic store of the spring in the watch; thermal store of the water (and surroundings); elastic store of the bow and bow string [decreases]; kinetic store of the arrow [increases]

4 a 500 (J) b 2.5 (kJ)

Page 115

1

Energy store	Equation
gravitational potential energy store *	$\Delta Q = m \times c \times \Delta\theta$
kinetic energy store *	$E = \frac{1}{2} \times k \times x^2$
elastic potential energy store	$Q = m \times L$
energy change during a change of state (e.g. melting)	$KE = \frac{1}{2} \times m \times v^2$
energy change when an object changes temperature	$\Delta GPE = m \times g \times \Delta h$

2

Select the correct energy equation	$KE = \frac{1}{2} \times m \times v^2$
Substitute in the correct values from the question using your highlighted values	$KE = 0.5 \times 3.0 \times (2.0 - 0)^2$
Calculate the answer	$KE = 6.0$
Choose the correct number of significant figures and add the units	$KE = 6.0$ J

Page 116

1 a Circled: 200 N/m and 0.50 m; underlined: 0.90

b $E = \frac{1}{2} \times k \times x^2$

c $E = \frac{1}{2} \times k \times x^2 = 0.5 \times 200 \times 0.502 = 25$ J

d Method 1

$$\text{efficiency} = \frac{\text{useful energy transferred by the device}}{\text{total energy supplied to the device}}$$

$$0.9 = \frac{\text{useful energy transferred by the device}}{2.5}$$

$$2.5 \times 0.9 = \frac{\text{useful energy transferred by the device}}{\cancel{2.5}} \times \cancel{2.5}$$

2.25 J = useful energy transferred by the device

Method 2

$$\text{efficiency} = \frac{\text{useful energy transferred by the device}}{\text{total energy supplied to the device}}$$

useful energy transferred by the device = total energy supplied to the device × efficiency

= 2.5 × 0.90 = 2.25 J

Page 117

1. $\Delta GPE = m \times g \times \Delta h$
2. $\Delta GPE = m \times g \times \Delta h = 0.15 \times 10 \times 2.5 = 3.75\,\text{J}$
3. As the apple falls, its gravitational potential energy decreases but its kinetic energy increases.
4. 3.75 J
5. $KE = \frac{1}{2} \times m \times v^2$
6. $KE = \frac{1}{2} \times m \times v^2$

   $v^2 = \frac{2 \times KE}{m} = \frac{2 \times 3.75}{0.15} = 50$
7. $v^2 = \sqrt{50}$, $v = 7.1\,\text{m/s}$ (to 2 sf)

Page 118

1. Gravitational potential energy $\Delta GPE = m \times g \times h$ $= 1.0 \times 10 \times 3.6 = 36\,\text{J}$
2. The student has written v instead of v^2.
3. $v = \sqrt{\frac{2 \times KE}{m}} = \sqrt{\frac{2 \times 36}{1.0}} = 8.5\,\text{m/s}$
4. **a** 27 000 J **b** $\frac{\Delta Q}{mc}$ **c** °C **d** 12.9 °C

Page 119

1 (a) There is a transfer from the elastic store of the spring (elastic potential energy) **(1)** to the kinetic store of the toy (kinetic energy) **(1)** and then to the gravitational store (gravitational potential energy) **(1)** of the toy.

(b) $E = \frac{1}{2} \times k \times x^2$ **(1)** $= 0.5 \times 50.0 \times (0.10)^2$

$= 0.25\,\text{J}$ **(1)**

(c) $\Delta GPE = m \times g \times \Delta h$ **(1)**; maximum height at efficiency of 1 = 0.42 m **(1)**; actual height at efficiency of 0.9 = 0.42 × 0.9 = 0.38 m **(1)**

Page 120

1 (a) $\Delta GPE = m \times g \times \Delta h$ **(1)** $= 0.40 \times 10 \times 2.0$ $= 8.0\,\text{J}$ **(1)**

(b) $KE = \frac{1}{2} \times m \times v^2$ **(1)**, $\sqrt{\frac{2 \times KE}{m}}$ **(1)** = 6.3 m/s **(1)** (Allow error carried forwards from (a).)

(c) As the ball hits the ground it deforms so kinetic → elastic **(1)**; as the ball rebounds elastic → kinetic **(1)**.

(d) Calculation of gravitational potential energy after bounce (5.6 J) **(1)**, use of efficiency =

$$\frac{\text{useful energy transferred by the device}}{\text{total energy supplied to the device}} \textbf{ (1)}$$

to give efficiency of 0.70 **(1)**. (Allow use of ratio of heights to reach efficiency.)

Unit 2

Page 122

1. A, e; B, f; C, a; D, g; E, d; F, h; G, c; H, b
2. **a** and **b**

A: support force, weight

B: support force, weight, compression, support force

3. **a**

drag

weight

b The weight would be the same but the drag (air resistance) would be larger.

Page 123

1. Quantity Symbol Unit

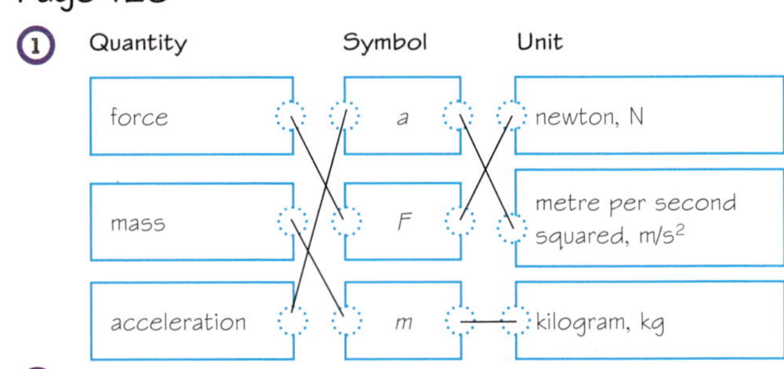

2.

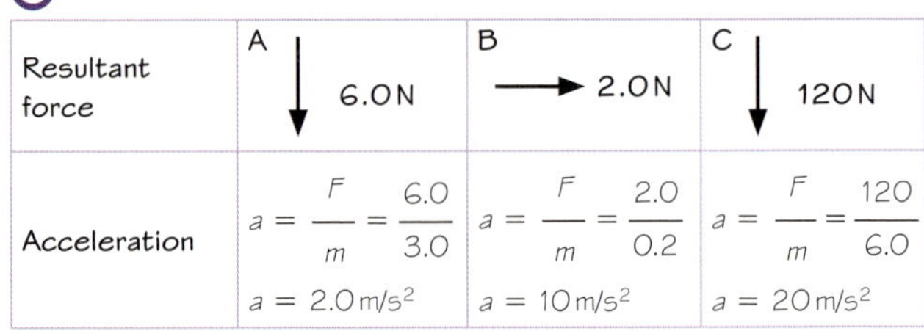

	A	B	C
Resultant force	↓ 6.0 N	→ 2.0 N	↓ 120 N
Acceleration	$a = \frac{F}{m} = \frac{6.0}{3.0}$ $a = 2.0\,\text{m/s}^2$	$a = \frac{F}{m} = \frac{2.0}{0.2}$ $a = 10\,\text{m/s}^2$	$a = \frac{F}{m} = \frac{120}{6.0}$ $a = 20\,\text{m/s}^2$

Page 124

1. first row: change in velocity = 4 m/s

   second row: end velocity = −3; change in velocity = −3 −9 = −12 m/s
2.

Stage	Your try
Scenario	A car has a mass of 1200 kg. Calculate the force needed to accelerate it from 0 m/s to 5.0 m/s in 5.0 s.
Find the change in velocity	0 to 5.0 m/s change in velocity = 5.0 − 0 = 5.0 m/s
Find the acceleration $a = \frac{\text{change in velocity}}{t}$	$a = \frac{\text{change in velocity}}{t}$ $= \frac{5.0}{5.0}$ $a = 1.0\,\text{m/s}$
Size of the forces involved $F = m \times a$	$F = m \times a$ $F = 1200 \times 1.0$ $F = 1200\,\text{N}$

Page 125

1 a A, B, D, E b A, D c A, D

2 b $p = m \times v$; $p = 5000 \times 8.5$; 42 500 (or 4.25×10^4) kg m/s to the right

3 $p = m \times v$; $4.0 \times 1.5 = 6.0$ to the right; $3.0 \times 2.0 = 6.0$ to the right; $6.0 + 6.0 = +12.0$ kg m/s to the right

Page 126

1 a $m = \frac{p}{v}$ b $\frac{0.3}{2.5} = 0.12$ kg

c The student has used too many significant figures. The data in the question has only two sf and so the answer should have the same number.

d The student has used the correct form of the equation ($p = m \times v$).

e The student has not found the change in velocity. They have subtracted the numbers but not taken into account the direction. The real change in velocity is 4.50 m/s. The student would get marks even though they used the wrong mass, as they did not score the mark for the calculation earlier.

f change in momentum = change in velocity × mass = $4.50 \times 0.12 = 0.54$ kg m/s

Page 127

1 (a) 0 m/s² **(1)**

(b) $p = m \times v = 0.045 \times 80 = 3.6$ **(1)** kg m/s **(1)**

(c) [–]3.6 kg m/s **(1)** (Allow error carried forward from part (b).)

(d) $a = \frac{\Delta v}{t} = \frac{(80-0)}{0.02} = 4000$ **(1)** m/s² **(1)**

(e) $F = m \times a = 0.045 \times 4000 = 180$ N **(1)** (Allow error carried forward from part (d).)

Page 128

1 (a) $a = \frac{\text{change in velocity}}{t} = \frac{(11-5)}{1.5} = 4.0$ **(1)** m/s² **(1)**

(b) $F = m \times a = 60 \times 4 = 240$ N **(1)**

(c) (–)240 N **(1)**

(d) $p = m \times v = 60 \times 5 = 300$ kg m/s **(1)** (accept N s)

Unit 3

Page 130

1 proton: positive; in the nucleus
neutron: neutral; in the nucleus
electron: negative; in orbit around the nucleus

2 a protons b protons; neutrons
c mass number; atomic number
d electrons; nucleus

3 a All of the chlorine isotopes have the same number of protons (17).
b All of the chlorine isotopes have different numbers of neutrons (18, 19 and 20).

4

Isotope and chemical symbol	Protons (from atomic number)	Neutrons (mass number – atomic number)	Electrons (same as protons)	Atomic notation
carbon-14 C	6	14 – 6 = 8 8	6	$^{14}_{6}C$
carbon-12 C	6	6	6	$^{12}_{6}C$
uranium-238 U	92	146	92	$^{238}_{92}U$
oxygen-17 O	8	9	8	$^{17}_{8}O$

Page 131

1 a protons b 2 c $^{4}_{2}He$

2 a negative charge b $^{0}_{-1}e$

3 a positive charge b $^{0}_{+1}e$

4

Radiation	α (alpha)	β⁻ (beta minus)	β⁺ (positron)	γ (gamma)
Symbol for radiation in equations	$^{4}_{2}He$	$^{0}_{-1}e$	$^{0}_{+1}e$	γ
Radiation consists of	two protons and two neutrons	a fast-moving electron ejected from the nucleus	a fast-moving positron ejected from the nucleus	electromagnetic radiation emitted from the nucleus

Page 132

1 $^{4}_{2}He$ 2 219, 4 3 219 = 215 + 4

4 86 = 84 + 2 5 $^{218}_{86}Rn$; $^{204}_{82}Pb$

6 79 = 80 + –1 7 $^{0}_{-1}e$; $^{14}_{7}N$

8

Decay type	Equation	Decay type	Equation
alpha	$^{185}_{79}Au \rightarrow ^{181}_{77}Ir + ^{4}_{2}He$	alpha	$^{231}_{91}Pa \rightarrow ^{227}_{89}Ac + ^{4}_{2}He$
beta	$^{14}_{6}C \rightarrow ^{14}_{7}N + ^{0}_{-1}C$	beta	$^{8}_{3}Li \rightarrow ^{8}_{4}Be + ^{0}_{-1}e$

Page 133

1 a 1800 Bq b 5 hours c 5 hours

2 $\frac{1}{16}$

3

Time in hours	0	→ 1st half-life	5	→ 2nd half-life	10	→ 3rd half-life	15
Activity in Bq	1800		900		450		225
Fraction remaining	$\frac{1}{1}$		$\frac{1}{2}$		$\frac{1}{4}$		$\frac{1}{8}$

Page 134

1 It means one half-life has passed.

2 a Four. There are four arrows.
b 16 000 Bq → 8000 Bq → 4000 Bq → 2000 Bq → 1000 (the answer is 1000 Bq)

3 a 90; 0; e b $^{90}_{38}Sr \rightarrow ^{90}_{39}Y + ^{0}_{-1}e$

Page 135

1 (a) Atoms of the same element (atoms with the same number of protons) (1) with different mass numbers/numbers of neutrons (1).

(b) Source A (1) (c) Source B (1)

(d) 4.5 days (1)

(e) $^{131}_{53}\text{I} \rightarrow {}^{131}_{54}\text{Xe} + {}^{0}_{-1}\text{e}$

(1 mark for 131, 1 mark for 54 and 1 mark for −1)

Page 136

1 (a) $^{226}_{88}\text{Ra} \rightarrow {}^{222}_{86}\text{Rn} + {}^{4}_{2}\text{He}$

(1 mark for 222, 1 mark for 86 and 1 mark for the whole helium nucleus)

(b) (i) The time it takes for the activity of a sample to fall to half of its original value **(1)**. (Accept half of mass/sample/count rate.)

(ii) One-eighth/$\frac{1}{8}$ **(1)**

Unit 4

Page 138

1 a, b, c

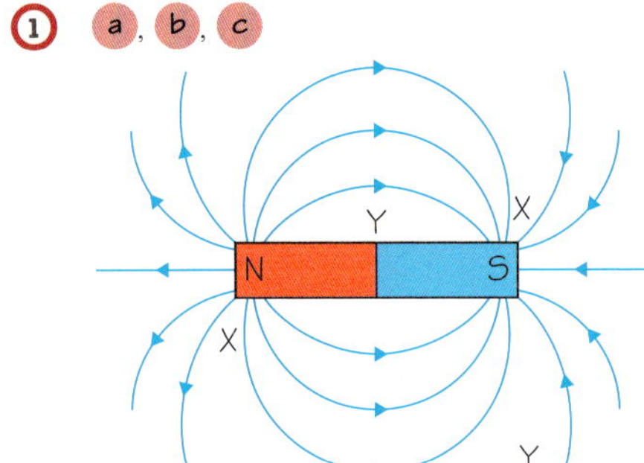

Other places for X and Y can be chosen. X should be where the magnetic field lines are close together and Y should be where they are far apart.

2

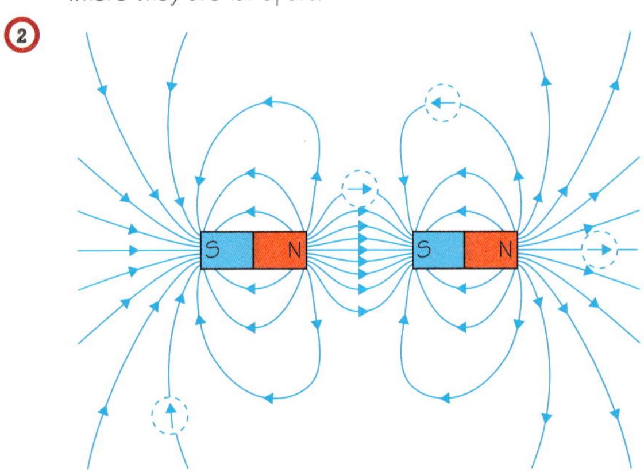

3 strong; uniform

Page 139

1 current; magnetic field

2 The current is moving upwards.

3 strongest; weaker; stronger

4 three upward arrows, one on each wire

5 True; False; True

Page 140

1 a, b, c

force on wire

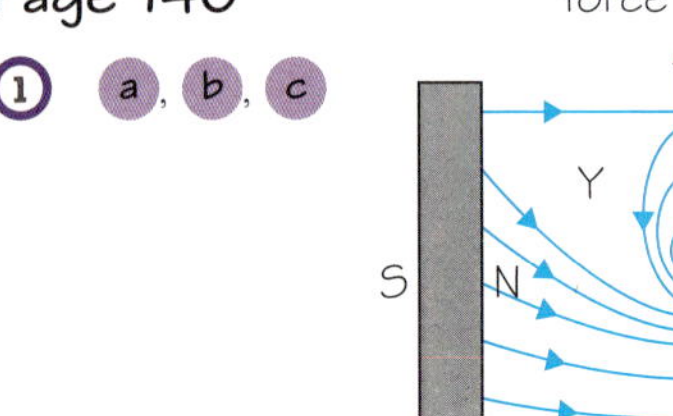

d opposite; the same as

2 left: up middle: left right: no force

Page 141

1

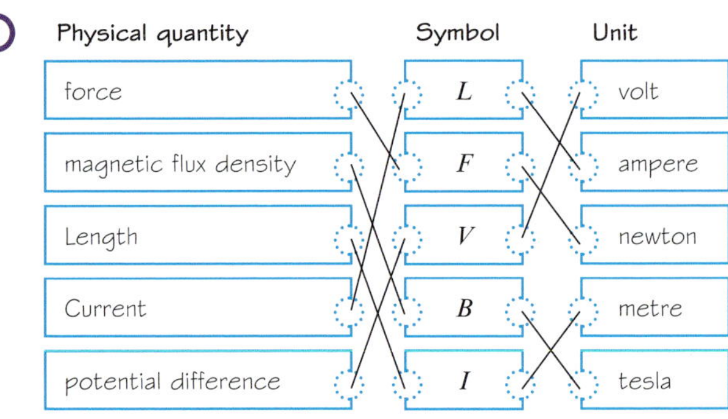

2 a magnetic field is pointing upwards

b force on the wire shown by an arrow pointing to the right

c force on the yoke is an arrow pointing to the left

3 $F = B \times I \times l = 0.2 \times 1.5 \times 0.12 = 0.036$ N

Page 142

1 a

b You should explain that the spacing between circles increases with distance from the wire because the magnetic flux density decreases with distance from the wire. You should explain that the field lines need arrows to show the direction of the magnetic field. In this case, the field is anticlockwise as predicted by the right-hand grip rule.

2 a The letter X would have been better in the middle of the solenoid where the field lines are closest together and not diverging.

b Other places of low magnetic flux density are shown on the diagram:

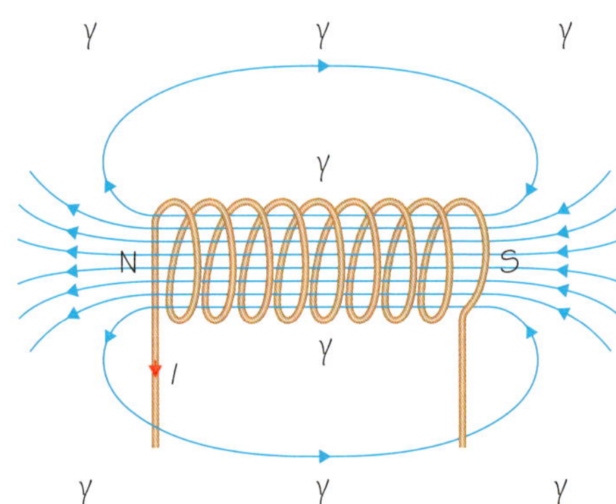

c The magnetic field is **strong** because the lines are close together and **uniform** because the lines are straight, parallel and evenly spaced.

Page 143

1 (a) A force acts on the wire because the current in the wire and the magnetic field from the magnets are at right angles to each other. **(1)** The magnetic field around the wire interacts with the magnetic field between the magnets **(1)** to create a force between the wire and magnets. **(1)**

(b) There should be an upward arrow on the wire between X and Y (The battery shows that the current is moving around the circuit in a clockwise direction. Fleming's left-hand rule allows us to predict the direction of the force on the wire.)

(c) force on a conductor at right angles to a magnetic field carrying a current = magnetic flux density × current × length ($F = B \times I \times l$)

(d) increase the magnetic flux density (allow magnetic field strength) **(1)**; increase the current; increase the length of wire in the magnetic field **(1)**

Page 144

1 (a) $F = m \times g = 0.016 \times 10$ **(1)**

$F = 0.16$ **(1)** N **(1)**

(b) $F = B \times I \times l$ **(1)**

$B = \frac{F}{I} \times l = \frac{0.16}{3.2} \times 0.1$ **(1)**

$B = 0.5$ **(1)** T **(1)**

(c) (i) The balance reading would change by more than 16.0 g.

(ii) Placing the magnets closer together would increase the magnetic flux density, B, **(1)** which would increase the force of the wire acting on the yoke. **(1)**

Unit 5

Page 146

1 P represents the physical quantities of pressure and power.

V represents the physical quantities of volume, potential difference, and the unit, volt.

2

Distance/length metre	Time second	Energy joule
Power watt	Mass kilogram	Current ampere
Force newton	Potential difference volt	Pressure pascal

3 kg m/s (the unit of mass multiplied by the unit of velocity)

4 $a = \frac{F}{m}$

Page 147

1 a and b

A lamp has a current of 1.6 A flowing through it.

Calculate the charge that passes through the lamp in 25 s.

c I, Q, t d $Q = I \times t$ circled

e $Q = I \times t = 1.6 \times 25 = 40$ C

2 a $I = \frac{P}{V}$

b $I = \frac{P}{V} = \frac{2.4}{12} = 0.2$ A

Page 148

1 a volume = $0.1 \times 0.1 \times 0\ 1 = 0.001\ m^3$

b $D = \frac{m}{V} = \frac{2.7}{0.001} = \frac{2700\ kg}{m^3}$

2

Prefix name and symbol	Multiplier	Multiplier (standard form)
mega (M)	1 000 000	10^6
kilo (k)	1000	10^3
centi (c)	$\frac{1}{100}$	10^{-2}
milli (m)	$\frac{1}{1000}$	10^{-3}
micro (μ)	$\frac{1}{1\,000\,000}$	10^{-6}
nano (n)	$\frac{1}{1\,000\,000\,000}$	10^{-9}

3

145 m = 0.145 km	145 m = 14 500 cm	2440 mm = 2.44 m
97.7 MHz = 97 700 000 Hz	48 mV = 0.048 V	101 300 Pa = 101.3 kPa
2300 W = 2.3 kW		

4

Object	Distance	Time	Average speed
car	240 m	40 s	6 m/s
lizard	300 cm	15 s	20 cm/s
rocket	480 km	64 s	7.5 km/s

Page 149

1 B Choose the right equation and write it down.

E Calculate the answer and give the unit.

A Identify the physical quantities, making sure the units are SI units if needed.

D Put the numbers in.

C Rearrange the equation if needed.

2

Step	Calculation
Identify the physical quantities, making sure the units are SI units if needed	$\Delta Q = 6.25$ kJ = 6250 J; change in temperature ($\Delta\theta$) = 16 °C, specific heat capacity (c) = ?
Choose the right equation and write it down	$\Delta Q = m \times c \times \Delta\theta$
Rearrange the equation if needed	$c = \frac{\Delta Q}{m \times \Delta\theta}$
Put the numbers in	$c = \frac{6250}{1} \times 16$
Calculate the answer and give the unit	c = 391 J/kg °C

3 $u = 18$ m/s; $v = 28$ m/s; distance (s) = 1.15 km = 1150 m; acceleration = ?

$v^2 - u^2 = 2as$

$a = \frac{(v^2 - u^2)}{2s} = \frac{(28^2 - 18^2)}{(2 \times 1150)}$

$a = \frac{(784 - 324)}{2300} = 0.2\ m/s^2$

Page 150

1 a 0.50 kg; 50 W; 45 °C b 3 minutes

2 a Did not convert the time in minutes to seconds. Gave the answer without a unit.

b $E = P \times t = 50 \times (3 \times 60)$

$E = 9000\ J = \Delta Q$

c Although the student used the incorrect value for ΔQ, they chose the correct equation, rearranged it correctly, and calculated the answer correctly.

d $\Delta Q = m \times c \times \Delta\theta$

$c = \frac{\Delta Q}{m\,\Delta\theta} = \frac{9000}{0.5 \times 45} = 400$ J/kg °C

Page 151

1 (a) $E = I \times V \times t = 4 \times 8 \times 300$

$E = 9600$ J

(b) $\Delta Q = m \times L$

$L = \frac{\Delta Q}{m} = \frac{9600}{0.0343} = \frac{280\,000\ J}{kg}$

Page 152

1 (a) $P = I^2 R$

(b) $P = 5000^2 \times 2.7 = 25\,000\,000 \times 2.7$

$P = 67\,500\,000\,\text{W} = 67.5\,\text{MW}$

(c) $P = 2500^2 \times 2.7 = 6\,250\,000 \times 2.7$

$P = 16\,875\,000 = 19.9\,\text{MW}$

The power loss in the second power line is one-quarter of the loss in the first power line.

Unit 6

Page 154

1

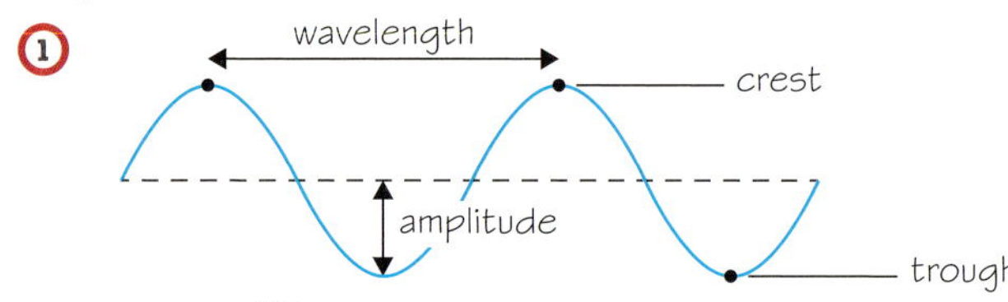

2 $v = \frac{x}{t} = \frac{0.5}{0.8} = 0.63\,\text{m/s}$

3 $v = f \times \lambda = 165 \times 2.0 = 330\,\text{m/s}$

4 a true b true c true d true e false

Page 155

1 incident; refracted; reflected; absorbed

2 B, D

3 reflected; refracted; absorbed; transmitted

Page 156

1 a true b true c false d true e false

2

	Gamma ray	X-ray	Ultra-violet	Visible light	Infrared	Micro-wave	Radio wave
filament lamp				✓	✓		
mobile phone						✓	(✓)
neon light				✓			
radioactive tracer	✓						

Note that satellite TV transmitters use microwaves, terrestrial (Earth-bound) TV transmitters use radio waves. Most modern mobile phones use microwaves, but some use radio waves.

3 refracted; reflected; absorbed

Page 157

1 Deep water:

$\lambda = \frac{v}{f} = \frac{0.50}{2} = 0.25\,\text{m}$

Shallow water:

$\lambda = \frac{v}{f} = \frac{0.40}{2} = 0.20\,\text{m}$

2

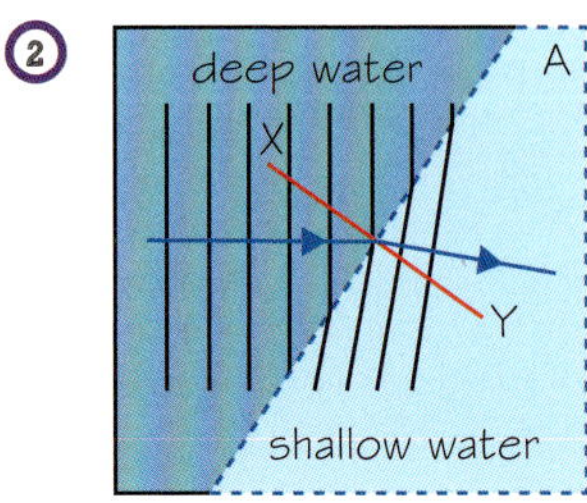

3 a increase; normal

b and c

shallow water B X Y deep water

4 a true b false c true d true

Page 158

1 The amplitude should be measured from a crest or trough to the centre of the wave.

The wavelength has been measured from the top of the crest to the top of the next crest, but not accurately enough.

2 There are eleven crests so there are ten waves. The student has calculated the answer with eleven waves instead of ten.

3 a Stating the correct equation and putting the numbers in (even though the answer for the wavelength was calculated incorrectly in part (b)). Calculating the answer correctly using those numbers.

b No mark was given for the unit because the speed would be calculated in cm/s not m/s.

Page 159

1 (a) normal line

(b) The wavelength decreases.

The frequency stays the same.

The wave speed decreases.

(c) Some energy from the light ray is reflected away from the block at A.

Some energy from the light ray is refracted inside the block at B.

Some energy from the light ray is absorbed by the glass.

Page 160

1 (a) Some energy is absorbed by the glass, the rest is transmitted by the glass.

(b)

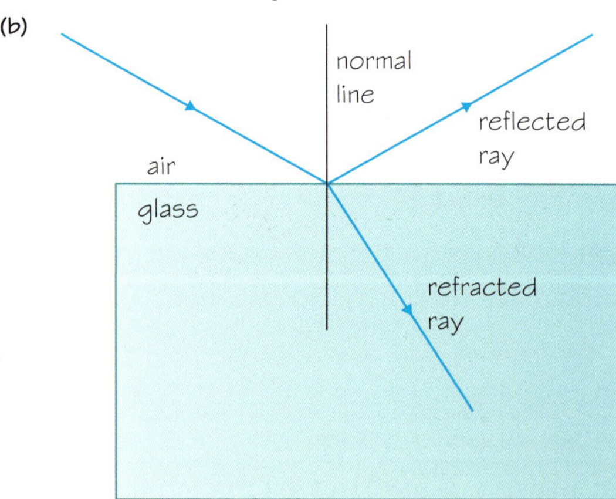

(c) Snow absorbs very little / reflects most of the light energy from the Sun making it very bright to look at; the sunglasses or goggles absorb most of the energy from the reflected light making it much less bright for the skier.

Unit 7

Page 162

Question parts

- Explain why this circuit could be used to measure resistance …
- Evaluate the use of radioisotopes such as iodine-131 …
- Describe an investigation a student …
- Compare and contrast the properties of water waves and sound waves.
- The current in the wire starts to slowly decrease when the circuit is left on. Suggest an explanation for why this reading changes.

Meaning

- Apply your knowledge and understanding to a new situation.
- Give similarities and differences between several things, not just one.
- Look at the information in the question and bring it together to make a decision and come to a conclusion with evidence from the question.
- Give an account of something, or link facts, information, events or processes in a logical order.
- Say how or why something happens – 'because' will be an important part of your answer.

Style of answer

- The current in the wire is causing it to heat up and so its resistance is increasing.
- Both water waves and sound waves need a medium to travel through. Water waves are transverse and sound waves are longitudinal.
- This circuit can be used to measure resistance because …
- Radioisotopes, such as iodine-131, may emit harmful gamma radiation but …
- Record the readings in a clear table and then plot a graph.

2 1, 4, 5, 2, 6, 3

Page 163

 Explain. Say how or why something happens.

b A student placed two magnets with opposite poles facing on a top-pan balance. She zeroed the balance.

She clamped a horizontal wire between the poles of the magnets, so the wire could not move.

She passed an electric current through the wire.

She observed that the top-pan balance reading changes.

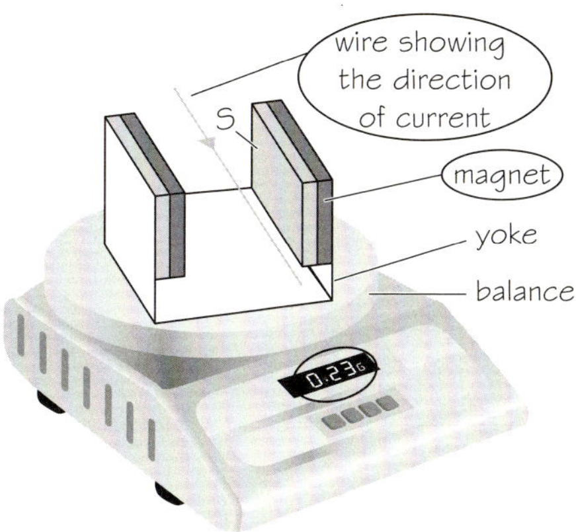

c why a force acts on the balance; how the force can be changed

2 A, b; B, d; C, a; D, c; E, f; F, g; G, e

Page 164

1 The motor effect

2 7, 5, 11, 4, 6, 9, 3, 8, 2, 10, 1

3 There is a magnetic field created around the wire because there is an electric current in the wire.

There is a uniform magnetic field created between the magnets because the magnets have opposite poles facing. A force acts on the wire because the magnetic field from the wire interacts with the magnetic field from the magnets.

There is a force acting downwards on the magnets because there is a force acting upwards on the wire from Fleming's left-hand rule.

The force increases if the current in the wire increases because the force exerted by the magnetic field on the wire is directly proportional to the current in the wire.

The force increases if the magnets are placed closer together because the strength of the magnetic field increases and the force acting on the wire is directly proportional to the strength of the magnetic field.

The force changes direction if the current in the wire changes direction because the direction of the force depends on the direction of the current and the magnetic field. It can be predicted by Fleming's left-hand rule.

Page 165

1 a air-glass boundary; inside the glass block; glass-air boundary

b reflection; refraction; absorption; transmission

2 Reflection and refraction; Absorption and emission of wave energy; The law of conservation of energy

3 a

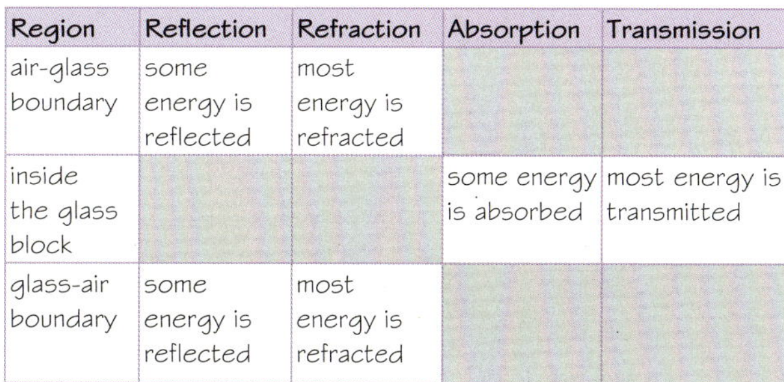

Region	Reflection	Refraction	Absorption	Transmission
air-glass boundary	some energy is reflected	most energy is refracted		
inside the glass block			some energy is absorbed	most energy is transmitted
glass-air boundary	some energy is reflected	most energy is refracted		

b As the ray of light passes through the glass block, some light energy is absorbed but the remaining light energy is transmitted by the block because all energy from the ray inside the block must be absorbed or transmitted.

As the ray of light reaches the glass-air boundary, some light energy is reflected internally but the remaining light energy is refracted at the surface and leaves the block because all energy from the ray must be refracted or reflected.

Page 166

1. These ideas are **not** needed: Which materials are good electrical conductors and which are insulators; The factors which affect the size of the force on a current-carrying wire; Which materials are magnetic and which are not.

2. This information is **not** relevant to the question: Copper is a good conductor but is not normally magnetic; If you make the current in the rod larger then the reading on the balance will change more because there will be an even larger force because of the equation $F = B \times I \times l$.

3. a This magnetic rod is then affecting the magnets because there is a force whenever two magnets are placed near to each other because of their magnetic fields.

   The magnets act to repel each other so the rod is pushed away from the permanent magnets.

   b When magnets put a force on the rod it is the same force that pushes back on the magnets and then the top-pan balance.

4. The current in the rod creates a circular magnetic field around it. This field interacts with the field from the magnets to produce a pair of forces acting in opposite directions on the rod and the magnets. The direction of the force from the frame on the rod is determined using Fleming's left-hand rule and is downwards. The rod is prevented from moving as it is clamped, so it exerts an equal and opposite upward force on the frame holding the magnets. There is a reduced resultant force on the balance from the frame, which decreases the balance reading.

Page 167

1. Explain
2. very high potential differences are used to transfer electrical power in overhead power cables; lower potential differences are used in houses
3. To transmit a large power, the current must be high or the voltage must be high (or both).
4. Use low resistances or low currents (or both).
5. A high voltage will allow a large power without a large heating effect in the overhead cable.
6. Explain power losses in cables using $P = I^2 \times R$.

   Need for transmission at high voltage.

   Need for lower voltage in homes for safety reasons.

7. Answer could include the following points in a logical order for 6 marks:

   The overhead power cables waste power because of electrical heating. To reduce this waste, a high voltage can be used because this allows the same power to be transmitted with a much lower current (from $P = I \times V$). This lower current means there is a lot less power wasted due to electrical heating in the cable (from $P = I^2 \times R$).

   High potential differences are very dangerous as they can cause a current to pass through you and electrocute you from a distance. A much lower p.d. of 230 V is a lot safer.

Page 168

1 Answer could include the following points in a logical order for 6 marks:

Both electromagnetic waves and sound waves transfer energy from the wave source to the surroundings. Both can be reflected or absorbed by a surface. Both electromagnetic waves and sound waves can be refracted at a boundary if they change speed. They both have different wavelengths and frequencies, which are related by wave speed = frequency × wavelength.

Electromagnetic waves can travel through a vacuum, but sound waves need a material to travel through. Electromagnetic waves are transverse waves, but sound waves are longitudinal waves. Electromagnetic waves are very fast, but sound waves are about a million times slower. Some electromagnetic waves can cause ionisation, but sound waves cannot.